Longman exam practice kit

GCSE Physics

Keith Palfreyman

LONGMAN

Series Editors
Geoff Black and Stuart Wall

Titles available
GCSE
Biology
Business Studies
Geography
Mathematics
Physics
Science

A-level
Biology
Business Studies
Chemistry
Mathematics
Psychology
Sociology

Addison Wesley Longman Ltd.,
Edinburgh Gate, Harlow,
Essex CM20 2JE, England
and Associated Companies throughout the World.

First Published 1997

ISBN 0-582-30379-6

British Library Cataloguing-in-Publication Data
A catalogue record for this book is available from the British Library.

Printed in Great Britain by Henry Ling Ltd.,
at the Dorset Press, Dorchester, Dorset

Contents

Acknowledgements

I am grateful to the following examination boards for their assistance:

- ▶ Midland Examination Group (MEG)
- ▶ Northern Ireland Council for Curriculum, Examinations and Assessment (NICCEA)
- ▶ Northern Examinations and Assessment Board (NEAB)
- ▶ Southern Examining Group (SEG)
- ▶ University of London Examinations and Assessment Council (London(EdExcel))
- ▶ Welsh Joint Education Committee (WJEC).

Where particular permission to reproduce questions has been granted the above initials will appear in the text. Please note that the answers given are those supplied by the author. In the case of questions supplied by an examination board the answers and hints are still those of the author and the board accepts no responsibility whatsoever for the accuracy or method of working in the answers that are given. Indications of mark schemes have not been approved by the examination board.

I am again grateful to my wife for more patience during the preparation of this book and for her constructive comments on my spelling and grammar. Thanks again to all those who willingly, or inadvertently in their examination answers, have also helped with the content of this book.

Keith Palfreyman

Introduction

This book is designed to help you improve your grade in your GCSE Science: Physics examination or to help you with the Physics part of the GCSE: Science double award. Each examination board sets its own syllabus based on the Physics content of the National Curriculum together with some extension work. This covers the work which is on all the examination syllabuses, plus the most common extension work. It is split into four parts.

Part I Preparing for the examination

This section covers what you should be doing before and during the examination. It will help you to plan your revision, remind you of the things that you should be doing before the actual day, and help you to understand what is required by the actual questions by looking at key words. A revision planner is provided to help you to structure your revision beginning some weeks before the examination. There are also important hints on how to answer examination questions.

Part II Revision topics and examination questions

This section has been split up into nine key areas. These are not all equal in length or difficulty but are chosen to keep similar material together and make your revision easier. In each area you will find:

▶ **Short revision notes** These will not replace your own full notes but will give most of the basic facts that you need and give you material in short fact form that is easier to revise. Important equations are repeated at the end of this section so that they are easy to find and check. If you still need more detail the Longman's Revise Guide for GCSE Key Stage 4 Physics is a good source.
▶ **A quick revision activity** This consists of short questions to test your knowledge of the facts covered in the section. It will tell you whether you need to spend more time revising that topic. The questions are all short and vary from filling in missing words to remembering equations or doing short calculations. The type of question depends on the topic.
▶ **Practice examination questions** At the end of each topic there are examination questions designed to test your knowledge of that particular topic. The questions are laid out in the same style as that of most examination boards. Do make sure that you try hard to answer questions to the best of your ability *before* you look up answers in the next section. Don't give up too easily or you aren't making the best use of the questions!

Part III All the answers

Here you will find all the answers to the revision activities and examination questions. The working for all of the calculations is clearly shown and full answers given so that you can see what the examiner expects, rather than only knowing the numerical answer at the end. The marks allowed for each part of the question are clearly indicated so that you can check your progress and make sure that you do not lose marks by not writing important detail when it is needed. Hints on avoiding common mistakes are also included in the answers.

Part IV Timed papers

This section contains two examination papers that cover a variety of topics in the same way as your final examination. The second one is aimed mainly at those who will do the higher paper. You will be able to allow yourself the time that is stated at the start of the paper, and then check your progress by marking the paper using the answers and marking scheme provided.

part I
Preparing for the examination

Planning for the examination

HOW TO USE THIS BOOK

Remember that the book and its examples are only part of your revision process. Read the revision sections of this chapter and make sure that you are planning your work properly. Each section has fairly comprehensive notes on the syllabus content, in most cases it is put into fact statements for you. Those sections that cause most problems in examinations are explained in a little more detail. Take care to get a proper copy of your syllabus and revise the material that *you* need. It is difficult to be exact because of all the different syllabuses, but the material that is usually only on the higher paper is marked **.

The examination questions are designed to test your knowledge of the parts of the syllabus that commonly occur in examinations. They are spaced out in the same way as they would be in the actual examination so that you can get used to judging the length of answer required and the number of facts for which there are likely to be marks. Most of the examination boards are no longer going to use any multiple choice questions in their new schemes. In spite of this I have included a few in each section of questions because they are useful as a quick test of your factual knowledge.

When you do a multiple choice question go through all the alternatives and make sure that you can see why the other four answers are incorrect – sometimes this elimination process is a good way to do the question or to check that you have done it correctly. You do not need to show working for this type of question but there is nothing to stop you using the space round the question to do calculations. In most cases the examination boards now use short answer questions of a variety of types. Use the space allowed and the marks allocated to judge whether you need to show detailed working – if in doubt write it on the answer paper. NEVER put your working for these, or the longer questions, on anything else other than the space provided on the answer paper.

The marks are given for each answer and you can use the questions at the end of each topic either as an aid to revision, doing the separate questions at different times, or as a test where you give yourself about an hour to see how many marks you can get. Do remember that the easiest questions are the ones that *you* can do and not necessarily the ones that come first.

In the real exam the markers use a system called 'error carried forward' to mark calculations. The idea is that you only lose marks once for each error. If you get a calculation wrong and then have to use its answer to do a second calculation, the marker will go through your answer giving credit for the method and science that you use, based on your first (incorrect) answer. In this way you can still get full marks for knowing how to do the second calculation.

PLANNING YOUR REVISION

► **Always start your revision in good time** It is never too early to start. Spreading the revision over a longer period of time will make it easier because you can do it in shorter sections and you can also plan a reasonable amount of leisure time while you are doing it.

- ► **Know your syllabus** If you are not sure which syllabus you are doing then ask a teacher and find out. Get a copy of it if you can. Mark off the sections that you have already covered – this usually helps to make you more confident.
- ► **Know about your examination** Make sure that you know how long the papers will be and what sort of questions to expect – will they be short, structured or longer and need fuller explanations? Which parts of the syllabus will appear on each examination – do you need to learn all of the material for each examination or only some of it?
- ► **Don't let the work build up** Be realistic about what you can achieve, make a revision plan and stick to it. During the course you should make sure that you have a full set of notes at the end of each unit and that all the homeworks are complete and marked so that they can be used to help you later.
- ► **Get help if you need it** Your teachers will always be pleased if you show interest by asking for information. Your parents or brothers and sisters may be able to help with some things and if they haven't done the sort of work in this course you might be able to do some of the revision with friends from the same class.
- ► When you have checked that you have all the basic information you need to make a **plan**.
- ► This book contains a **revision planner** that assumes that you will begin to do serious revision at least twelve weeks before your examinations. There are full instructions on it and you should use it so that you can check your general progress as you approach the examinations. You will need to check the exact date of your own examinations so that you can work out the starting date. Closer to the examination you should find that you get 'official' printed timetables for your examinations from the examination boards where you have entries. These will come to you from your school and will also enable you to check that you have been entered for the correct level.

REVISION TIPS

- ► Plan ahead. Use the planner and when you have made a plan stick to it.
- ► Make sure that you have somewhere to study properly. You will need a quiet place at home where you won't be interrupted by other members of your family.
- ► Keep the revision sessions short for each subject. You cannot concentrate for long periods and you will probably be at your best for about forty minutes. After this, review your progress and stop or at least change to another subject.
- ► Practise answering questions from past papers or questions from other papers and books of about the correct level. If you are not doing the higher level paper then don't get discouraged if you can't do the higher level questions – you probably won't have been taught the additional material that is needed.
- ► Don't work late at night. It means that you are planning your time badly and will only result in tiredness that makes work harder the following day.
- ► Make sure that you plan some leisure and some time off into your schedule. You will feel better if you reward yourself for your hard work.
- ► Tick off the topics that you revise on a list as you go along. This will show you that you are making progress and help to boost your confidence.
- ► Don't keep stopping to make drinks or get something to eat. It usually means that you are concentrating badly and the work then takes longer. Use this sort of thing as a reward after *completed* sections of revision.
- ► Think, act and be *positive*. You *can* do it if you want to so don't talk yourself out of it.
- ► Try different ways to revise. It can be fun to revise with a friend, but make sure that you can keep the work going and the session doesn't just turn into a chat.
- ► If you start to feel under stress, talk to your parents or teachers about it – *before* keeping up with work becomes a serious problem.

► Find a revision process that works for you. Just reading notes is a failure for most people and they quickly lose concentration. Most people need an *active* process to keep them interested. Copying out notes is a very time consuming process and is not efficient. You need to DO something to reinforce your learning. Try one of the following:

– Write a keyword to remind you of each paragraph. After a few pages look back at the key words and see if you can remember the content of the paragraph.

– Practise writing short summaries of descriptions.

– Make lists of equations and definitions. These are short and factual and you can test yourself until you know them all. These lists can be fun to test if you make them into a competition game with a friend – who can get the most right in succession? Keep score and change over each time one of you gets one wrong.

– Re-do some of the calculations that you have been given for homework.

– Draw a spider diagram. This means putting the title of the topic in the middle of a sheet of paper and then writing the name of each part of the topic that you revise on to the sheet. Join the topic to the main title with a line. Some topics can be divided even smaller by further lines, or even link to each other. The resulting diagram is a good start for you to check what you can remember.

A part completed spider diagram

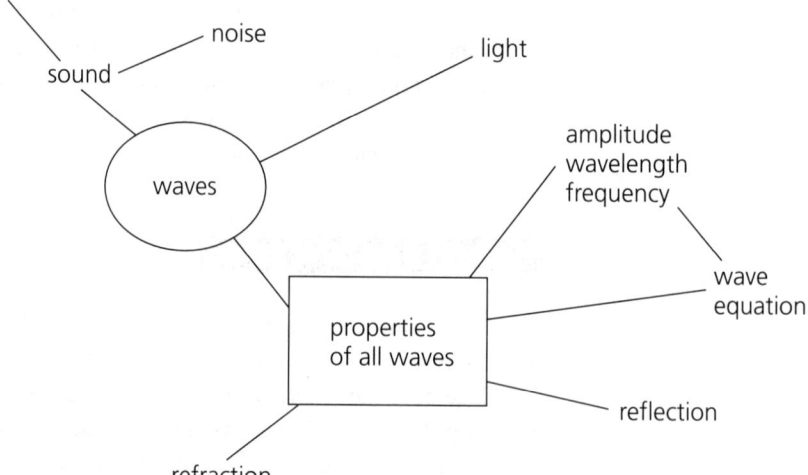

– If you have made a spider diagram, brief summary or a list of key words, go over them quickly again about a week later. It is a great help to your confidence as you start to find that you have remembered the important parts.

– There are many other ways that you can try – one of my students recorded all the important equations on to a cassette so that he could listen to them on his personal stereo on the school bus. Find methods that help you but remember that it will be best when you have to DO something as its helps you to concentrate.

BEFORE THE EXAMINATION

► Check that you have all the equipment that you will need: pens, pencils, rubber, ruler, calculator with a new battery.

► Don't buy a new calculator just before the examination and expect that you can learn to use it at the last minute. Do use a calculator that you are familiar with and that doesn't cause you to waste time or write down strange answers.

► Check the date and time of the examination. All the revision and several years of hard work can be wasted if you miss the examination!

▶ Make sure that you are going to be on time. Most examination boards will not accept missing a bus as a reasonable excuse and won't mark work done after the official ending time of the examination.

DURING THE EXAMINATION

▶ **Arrive fully prepared** This doesn't just mean having done the revision. Have you got a pen, pencil, rubber and ruler? Do you have spare ones in case one stops working? Have you remembered your calculator? Does it have a good battery in it?

▶ **Fill in the front part of the answer paper carefully**, including your candidate number and centre number. It is important that your paper can be correctly traced if it is mislaid later!

▶ **Read the instructions** on the front of the paper carefully. Make sure that you really do know how long the paper should take – don't assume that it will be the same as last year or the practice papers that you have been doing. Is there an information sheet with equations or other facts on it that will help you? Sometimes this is the very last page so don't discover that it is there when it is too late.

▶ **Check which questions you have to do** In this sort of examination you do not usually have a choice and you have to answer all the questions.

▶ **Share out the time properly** Even though you have to do all the questions there will probably be more than one section and the sections will contain different types of question. In the instructions on the front of the paper there will usually be a sentence that tells you not to spend more than a certain time on each section. This is important as the sections will not all have equal total marks.

▶ **Read the first question slowly** and begin to answer the paper. Don't begin in a rush and panic so that you make silly mistakes or miss out parts of questions that you can do.

▶ **Do have a go at *all* of the questions** There may be some marks that you can pick up by having a go even if you don't know enough to get them all. You can't do worse than leaving a blank space which will obviously get no marks at all!

▶ Make sure that you **answer the question that is asked** and not a similar one that you would rather have been asked. The marking scheme will be exact and aimed at only the facts that fit the question on the paper.

▶ **Do use all the help** that is built into the question paper for you. The size of the space that has been left for the answer will help you to decide how much writing you need to do and how much detail is needed. The number of marks for each part of a question is also in the margin. On a science paper this will usually be the same as the number of facts that you are expected to have in your answer. The marking scheme will then have one mark for each fact. If there are four marks available then try not to use all the space for one or two facts and have no chance of the other marks.

▶ If you are **doing a calculation** then **do** show the working. Many questions help you by asking for the equation first, but try to get into the habit of following this order:
1 write down the correct equation
2 substitute the numbers in the correct places
3 work out the answer that you need
4 remember to add the correct unit for your answer.
If you do this you will have shown the examiner that you understand the science involved and have not just guessed what to do with the numbers. If you make a mistake, with the arithmetic for example, you can still get most of the marks. If you only write down a final answer you will certainly lose *some* marks and you will lose *all* of the marks if you have made a mistake.

▶ **Check that you have fully completed each question before you leave it**. It is easy to answer one part successfully and then move on thinking that you have done it all.

▶ Try to leave a little time spare at the end so that you can **check the work that you have done.** Try to find the silly mistakes before the examiner does! Don't leave questions unattempted.

▶ **If you start to run out of time** be sensible and keep your answers clear and readable. Remember that it is usually easier to get the first few marks on a new question than to continue trying to get the last few harder marks on another question.

KEY WORDS

Some words have quite exact meanings in examinations and you need to make sure that you understand what the examiner will want you to do.

▶ **Calculate** Work out the value of something. **Do** show the working unless you are told that it is not needed. e.g. Calculate the density of the metal.

▶ **Compare** Are the things in the question alike or similar? What are the similarities and differences? e.g. Compare the iris of an eye with the diaphragm of a camera.

▶ **Complete** Fill in the spaces in a sentence or a table. e.g. Complete the following table showing the names of the planets.

▶ **Define** Give the *exact* meaning of something. e.g. Define *half-life*.

▶ **Describe** Write in detail about something. e.g. Describe what happens when you close the switch in the circuit.

▶ **Draw a diagram** Draw a clear *labelled* diagram. Use a pencil and ruler, this is *not* asking for a rough sketch. e.g. Draw a diagram of a simple thermometer.

▶ **Explain** Make it clear what is meant by something. e.g. Explain how air exerts a pressure.

▶ **Explain why** Give scientific reasons why something happens. Don't just describe what it is that happens – you are often told that in this sort of question anyway. e.g. Explain why the air rises in the syringe.

▶ **Name** Write down the name of something. You don't need to write about it at all, just write down the name. e.g. Name a gas that could be used to fill the balloon.

▶ **Plot** Put the exact points on a graph.

▶ **Predict** Use the information in the question to say what you think will happen.

▶ **State** Write down briefly the exact facts about something. It can mean the same as 'define' when it says 'state what is meant by'. e.g. State what is meant by *density*.

▶ **Suggest** Use your *scientific knowledge* to make *reasonable* suggestions. e.g. Suggest a use for the circuit shown.

▶ **What happens to** This usually just wants a short statement, a fact. e.g. 'What happens to the armature when the electromagnet is switched on?' will just want you to state that the armature is attracted to the electromagnet.

SYLLABUS CONTENT

Each of the examination boards sets its own syllabus. All the syllabuses have the same common core so that they meet the National Curriculum but the additional material can be different on each syllabus. This book revises all the most common material, but if you need more detail or some work on a topic that is only on one of the syllabuses then you should revise it in the Longman Revise Guide for GCSE Physics.

EXAMINATION BOARDS

If your school has not got spare copies you can order copies of your particular syllabus from the examination board at one of the following addresses:

AEB, Stag Hill House, GUILDFORD, Surrey GU2 5XJ
London (EdExcel), Stewart House, 32 Russell Square, LONDON WC1B 5DN
NEAB, Devas Street, MANCHESTER M15 6EX
NICCEA, Stranmillis College, Stranmillis Road, BELFAST BT9 5DY
WJEC, 245 Western Avenue, CARDIFF CF5 2YX

part II
Revision topics and examination questions

Static electricity and basic circuits

Topics marked ** are usually on the higher paper and may not be on all syllabuses.

TOPIC OUTLINE

Static electricity

▶ **Neutral objects** have equal numbers of positive and negative charges. The positive charges are on protons in the nucleus of the atoms and do not move. The negative charges are carried by electrons. These are at the outside of atoms and can be moved from one material to another. You can think of the earth as a large neutral object which is not changed by the loss or addition of a few charges.

▶ An object gets a **negative charge** when it gains some electrons.

▶ An object gets a **positive charge** when it loses some electrons.

▶ Objects can be charged by friction such as rubbing with fur or a cloth. Electrons are then transferred from one material to the other so that one becomes negative (gains electrons) and the other equally positive (loses the same number of electrons).
   – **Opposite charges attract**.
   – **Charges that are the same repel**.
   – Charged objects will also attract uncharged objects (e.g. a plastic pen rubbed on your sleeve will become negatively charged and can pick up small pieces of paper).

▶ Because charges that are the same repel each other they will spread out as far as they can. They will cross to earth if they can so that they are spread out as far as possible. This means that conductors such as metals cannot hold a static charge unless insulated from earth whereas a plastic, such as a polythene rod can hold charge as it cannot leak away through the insulator. Charged conductors can always be discharged by connecting them to earth.

▶ **Larger charges** on an object produce **greater voltages**. If the voltage becomes great enough a **spark** will jump between the object and earth (or any earthed object close to it). If the object is able to carry enough charge the current in the spark can be large enough to be able to give a dangerous electric shock. The spark can be dangerous if near to flammable gases.

▶ **Other examples** of charges that you might meet are charges on clothing ('sticks' to you, crackles, sparks), charges built up by walking on thick carpet, lightning.

▶ **Uses** include charging the droplets of paint and insecticide from sprays so that they spread out better and are attracted to earthed objects (cars and plants), smoke precipitators (a charged grid attracts the dust in the smoke from chimneys) and the attraction of ink to paper in photocopiers.

Basic circuits

▶ You should have met these basic ideas:
   – There must be a battery or a power supply to supply the electrons and drive them round the circuit. There must be a complete circuit for current to flow.

- The electric current will be a flow of charge, usually carried by electrons and sometimes by ions.
- You should also know the basic circuit symbols that are used internationally.

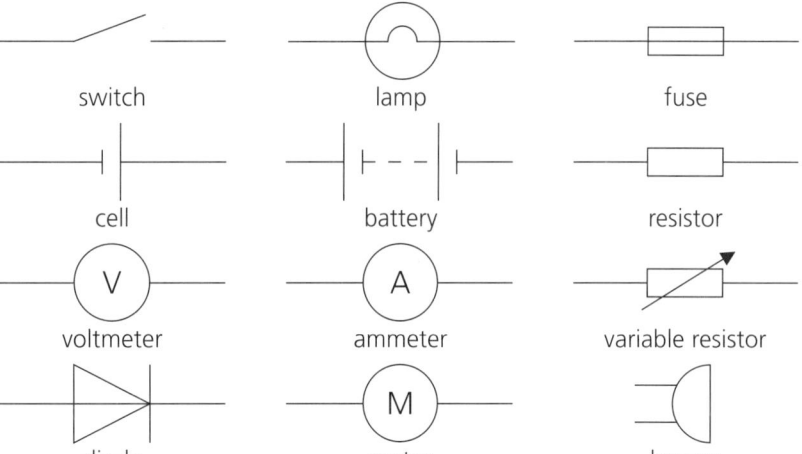

Circuit symbols

- A bigger potential difference (p.d.) across a component will drive a bigger current through it.
- P.d. is measured in **volts** by a **voltmeter** connected across the component.
- Current is measured in **amps** by an **ammeter** connected in series so that the current to be measured must flow through it.
- When components are connected **in series** (one after the other) the current through them all is the same and the total p.d. across them is shared between them. When **bulbs** are connected **in series** they will all go out if one of them fails and will only get to normal brightness if the total p.d. is the sum of all the individual p.d.s needed. **Cells in series** have their voltages added.
- When components are connected **in parallel** the current divides; part goes through each component and then it joins again. There will be the same p.d. across each component. The total current is the sum of the currents in each of the components. **Bulbs** are often connected **in parallel** so that if one fails the others are unaffected. **Cells** connected **in parallel** can produce more current if required to do so.
- **Direct current (d.c.)**, like that from batteries, always flows in the same direction round a circuit. **Alternating current (a.c.)** constantly changes direction. The a.c. from 'mains' flows once each way 50 times a second.

▶ **✷✷The charge** transferred by an electric current depends on the size of the current and the time that it is flowing for. Charge will be in coulombs (C) and the current must be in amps and the time in seconds. One coulomb will be transferred when one amp flows for one second.

▶ **Resistance** is a measure of how hard it is for current to get through a component. Long thin lengths of wire will have more resistance than short thick ones. The resistance also depends on the metal used.
✷✷You can also calculate resistance from the equation known as Ohm's law. The resistance will be in ohms (Ω) and the p.d. must be in volts and the current in amps.

▶ Resistances of components in series will add. E.g. putting more bulbs in series will create more resistance so that less current gets through and the bulbs are dimmer. Five bulbs of resistance 6 Ω each will have a total resistance of 30 Ω.

▶ **✷✷Resistors** in parallel will have a total resistance that you can find from the equation shown.
In the simplest case this will mean that two resistors of equal resistance connected in parallel will have a total resistance equal to half of one of them.

KEY POINT
Charge transferred = current $\times$ time

KEY POINT
Resistance =
$$\frac{potential\ difference}{current}$$

KEY POINT
$R_{total} = R_1 + R_2 + \dots$

KEY POINT
$$\frac{1}{R_{total}} = \frac{1}{R_1} + \frac{1}{R_2}$$

An example

Two resistors of resistance 5 Ω and 4 Ω are connected in parallel.
What is the total resistance?

$$\frac{1}{R_{total}} = \frac{1}{R_1} + \frac{1}{R_2}$$

$$\frac{1}{R_{total}} = \frac{1}{4} + \frac{1}{5} = \frac{5+4}{20} = \frac{9}{20}$$

$$R_{total} = \frac{20}{9} = 2.2\ \Omega$$

NOTE: you need to add the two fractions properly!

Remember that the total of resistances in parallel will always be smaller than the smallest of the individual resistances.

▶ You can assume that wires and resistors will obey Ohm's law and that their resistance will stay the same, but the law is only true if the temperature stays the same. Here are some **examples where the resistance changes**:
  – A bulb – as its filament gets hotter its resistance will increase.
  – A diode – current is only allowed through in one direction and the resistance is very large in the reverse direction.
  – A light dependent resistor has less resistance in brighter light.
  – A thermistor will have less resistance as its temperature increases.

> **KEY POINT**
> ***Energy transferred = voltage × charge transferred*

▶ **Energy** is released whenever current flows round a circuit and depends on the p.d. (voltage) and the charge that is transferred. If the energy is transferred in a resistor then it becomes heat.
  **Since charge is given by $Q = It$, the equation is the same as saying $\mathbf{E = VIt}$.
  Remember that energy will always be measured in joules.

> **KEY POINT**
> *Power (in watts) = potential difference (in volts) × current (in amps)*

▶ **Power** is the rate at which electrical energy is being transferred. A power of one watt (W) means that one joule of electrical energy is being transferred each second.

An example

A small bulb is marked 2 V, 0.25 A.
What is its resistance?

$$\text{Resistance} = \frac{\text{potential difference}}{\text{current}}$$

$$= \frac{2}{0.25} = 8\ \Omega$$

What is the power of the bulb?
Power = volts × amps
  = 2 × 0.25 = 0.5 W

How much energy is transferred into heat and light each minute?
0.5 W = 0.5 joules per second.
so energy transferred per minute
  = 60 × 0.5 = 30 J

How much charge is transferred each minute?
Charge transferred = current × time
  = 0.25 × 60
  = 15 C

▶ **Fuses** contain a thin wire that will heat up when current flows through it. If the current is above the level for which the fuse is designed, then the fuse will melt in the middle and break the circuit. Fuses are designed to prevent fire or damage to cables caused by overheating and are NOT there to prevent electric shocks. The smallest possible value of fuse should be used and placed in the wire between the + or live of the supply and the appliance. Mains circuits may have a circuit breaker instead of a fuse, which automatically switches off the circuit when a certain current is reached.

▶ An **earth wire** connects the metal cases of mains appliances to earth and is to prevent electric shocks. If the case is connected to the live wire of the supply by a fault, then current flows to earth through the earth wire instead of through the next person to touch the case. The earth connection will usually take so much current that the fuse is also blown. Appliances that are **double insulated** don't need an earth wire as they are made so that the live wire cannot touch outside metal parts. A special circuit breaker called an RCB can also be used instead of a fuse and switches the circuit off if current flows to earth.

▶ **Mains** electricity is supplied at about 230 V and is a.c.

The **neutral wire stays at close to zero (earth voltage) and the **live** wire changes between a negative and positive voltage compared to the neutral. The frequency is 50 Hz.

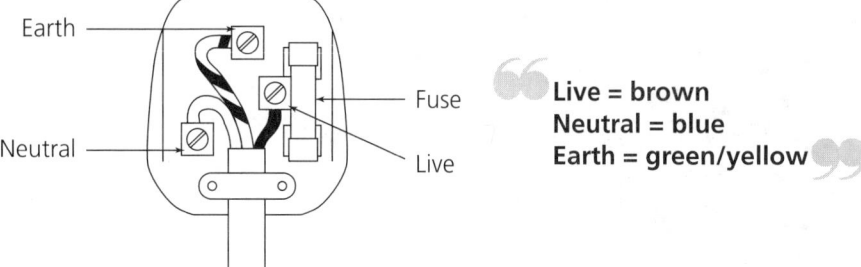

A three pin plug

> **Live = brown**
> **Neutral = blue**
> **Earth = green/yellow**

– Mains is supplied as a.c. because it can be changed to any voltage needed by a transformer.

– The energy used by an appliance can be worked out by using the equation shown.

The energy is measured in kilowatt hours because the joule is too small to be useful in this case. You can the work out the total cost if you know the cost of each kWh. One kWh is commonly called a unit.

Total cost = kWh × cost per unit

Remember that you must work in hours and that 1 kW = 1000 W.

KEY POINT
Energy in kWh =
power in kW × time in hours

Magnets

▶ You should already have met the following at key stage 3:

– The two areas on a magnet where the force is strongest are called **poles**.

– If the magnet is free to turn, it will point approximately north–south. The poles are then called the **north** (seeking) pole and the **south** (seeking) pole.

– **Like poles repel; unlike poles attract.**

– A **magnetic field** round a magnet produces a force on any other magnet or magnetic material placed in it. The field is strongest near the poles and gets weaker as distance from the magnet increases. The shape of the field can be shown by lines of force. The lines can be plotted using a plotting compass or shown by sprinkling fine iron filings over the area to be tested. You should be able to draw the field round a bar magnet.

▶ **Electromagnets** can be made by sending a current through a coil of wire.

– The field round the coil is the same shape as that round a bar magnet.

– The strength of the field produced can be increased by using more current, more turns on the coil, and by putting an iron core inside the coil.

– Reversing the direction of the current reverses the poles of the electromagnet. Using a.c. will reverse the poles constantly – but this will not matter in applications like picking up scrap which is attracted by either pole.

– Your earlier work should also enable you to explain how an electromagnet is used in an electric bell and a relay.

The motor effect

▶ A wire carrying a current in a magnetic field will have a force acting on it. The force can be made greater by:

– using a larger current

– using a stronger magnet.

► The direction of the force can be reversed by:
  – reversing the direction of the current
  – reversing the direction of the magnetic field.

► **The direction of the force is often found by using **Fleming's left hand rule** – hold the thumb, first and second finger of the left hand at right angles to each other. Point **F**irst finger in the direction of the **F**ield, se**C**ond finger in the direction of the **C**urrent and the thu**M**b will point in the direction of the **M**ovement produced.

► You can use your knowledge of this force and its direction to work out which way the force will act on each side of a coil in an electric motor.

► Loudspeakers also have a coil that carries a current in a magnetic field. The changing current produces a changing force on the coil and causes the cone attached to the coil to vibrate. The vibration of the cone produces the sound that you hear.

Useful equations

Charge transferred = current × time $Q = I \times t$
Voltage = current × resistance $V = I \times R$
Resistors in series add $R_{total} = R_1 + R_2 + \dots$

Resistors in parallel use the equation $\dfrac{1}{R_{total}} = \dfrac{1}{R_1} + \dfrac{1}{R_2} + \dots$

Energy transferred = voltage × charge transferred $E = V \times Q$
Power = potential difference × current $P = V \times I$
Energy in kWh = power in kW × time in hours

★ REVISION ACTIVITY

Make sure that you can do the following short checks and then attempt the main revision questions. Remember to write down any equations that you use.

1 Negative charges are carried by _____.

2 Larger charges produce _____ voltages on isolated objects. If the voltage is large enough a _____ may jump between the object and _____.

3 Like charges (charges that are the same) will _____ each other.

4 Bulbs connected one after the other are in _____ and the _____ through them all will be the same.

5 An electromagnet can be made stronger by _____ or _____.

6 If a bulb takes 2 A from a 12 V battery its resistance is ____Ω

7 The power of the bulb in question 6 is _____W

8 Three identical bulbs are connected in series. Each bulb has a resistance of 6 Ω. What is the total resistance?

9 If the same bulbs were connected in parallel, would the total resistance become more or less? _____ Would the total current supplied to the bulbs be more or less? _____

10 A 100 W light bulb was left on for 25 hours. How many units of electrical energy were used?
   What was the cost of leaving the bulb switched on if electricity costs 7 p per unit?

EXAMINATION QUESTIONS

Questions that are most likely to be on the higher paper only are marked **

Question 1
Which of the following is true for a three pin plug?
A The earth pin is connected to the brown wire
B The live pin is connected to the blue wire
C The neutral pin is connected to the brown wire
D The long pin is the neutral pin
E The earth wire is yellow and green

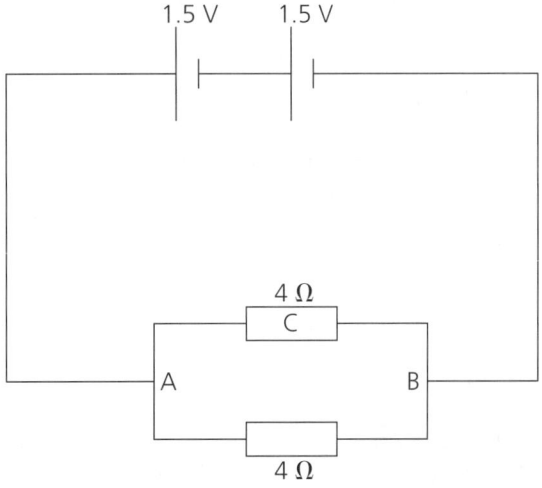

Question 2
(a) What is the voltage across AB?   [1]

....

(b) What is the current through resistor C?   [2]

....

....

....

Question 3

Electrical appliance	Power rating/kW	Power rating/W
Television	0.2	200
Electric kettle		2000
Food mixer	0.6	

The table above shows the power rating of three electrical appliances.
(a) Fill in the blank spaces in the table.   [2]
(b) State which appliance transfers the least amount of energy per second.

.... [1]

(c) State which appliance converts electrical energy into heat and kinetic energy.

.... [1]

WJEC 1996

Question 4

Read the following passage carefully before answering the questions.

Spraying crops with chemical fertilizers or insecticides has become more efficient. A portable high voltage generator gives the drops of liquid insecticide a small positive charge. This makes the liquid break up into smaller droplets and causes the spray to become finer and spread out more.

The plants, which are all reasonable conductors, are in contact with the earth. As the droplets of spray get near the plants, the plants themselves become slightly charged and attract the droplets.

(a) (i) Explain why the positive charge on the droplets makes the spray spread out. [1]

....

(ii) State what charge appears on the plants as the droplets come near to them. [1]

....

(b) Suggest **two** reasons why it is an advantage to both the farmer and the environment to use very small charged droplets during insecticide spraying. [2]

....

....

....

....

WJEC 1996

**Question 5

The figure below shows apparatus used to demonstrate the motor effect. **X** is a short length of bare copper wire resting on the other wires.

Horseshoe magnet
Bare copper wire
N
S
RED
BLACK
Power supply
Copper wire **X**

(a) (i) Describe in detail what happens to wire **X** when the current is switched on.

....

....

....

(ii) What difference do you notice if the following changes are made?
A The magnetic field is reversed.

....

....

B The current is increased.

....

.... [4]

(b) The figure below shows a coil between the poles of a magnet.
The arrows indicate the direction of the conventional current.

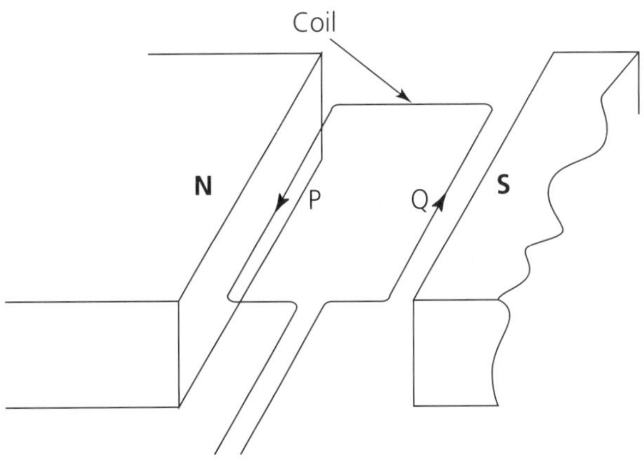

On the above figure draw arrows to show the direction of the force on:
(i) side P of the coil
(ii) side Q of the coil [2]

(c) Explain why these forces cause the coil to rotate.

....

.... [2]

(d) The figure below shows the coil in a vertical position.

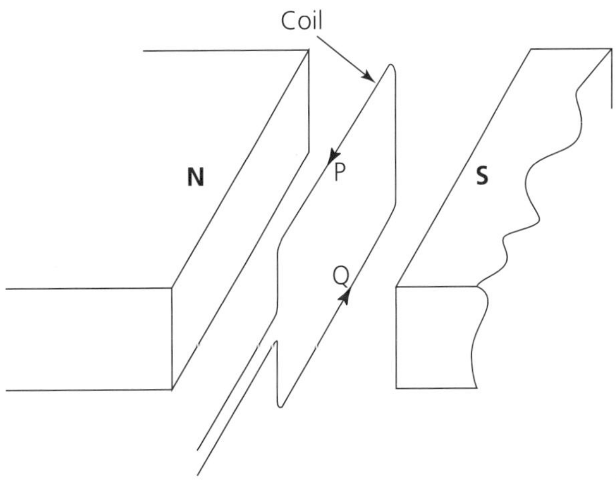

Draw arrows on the above figure to show the direction of the force on:
(i) side P of the coil
(ii) side Q of the coil [2]

(e) What will happen to the coil now?

....

.... [1]

(f) State **one** change which could be made which would make the coil rotate
continuously?

....

.... [1]

NEAB 1996

2 Electricity and electronics

Topics marked ** are usually on the higher paper and may not be on all syllabuses. The amount of electronics on the new Physics syllabuses varies enormously. Check with your syllabus so that you are not revising unnecessary material.

TOPIC OUTLINE

Electromagnetic induction

▶ **A voltage will be produced** when:
  – a wire is moved in a magnetic field or
  – the magnetic field through a coil changes.
  The **voltage can be made larger** by:
  – making the movement faster
  – using a stronger magnet
  – using a coil with more turns.
  The **voltage can be reversed** by:
  – reversing the direction of the movement
  – reversing the direction of the magnetic field.
  If the wire or coil is part of a complete circuit then the voltage will produce a current in that part of the circuit. You should be able to apply this rule to the coil in a simple a.c. generator, consisting of a coil rotating in a magnetic field, to show how it works.

▶ **Transformers** are used to step up or step down the voltages in a.c. circuits.

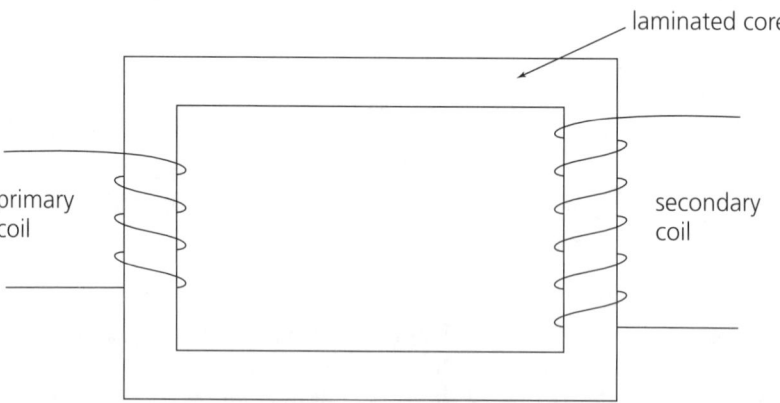

A transformer | primary coil | laminated core | secondary coil

**The changing current in the primary coil (note that this means a.c. is essential) makes a changing magnetic field in the core. The changing magnetic field in the core is through the secondary coil and induces a voltage in it. The core is laminated to avoid heat losses by currents induced in the core instead of in the coils.
**The input and output voltages are connected by the equation shown.

▶ Transformers are especially important in the transmission of power in the **National Grid**. The power is transmitted at low current to avoid heat losses in the cable and this means using high voltages. Transformers are used at each end of the cable to step the voltage up and then down to 230 V at the consumer end. Thick cables are also used as their low resistance also reduces heat loss.

KEY POINT

$$\frac{\text{voltage across secondary}}{\text{voltage across primary}} = \frac{\text{turns on secondary}}{\text{turns on primary}}$$

Electronics

▶ The following additional components are used:
  – The **LED** (Light Emitting Diode) behaves as a normal diode but also emits light when current flows through it. Remember that it will not light up if *reverse biased* (connected with the voltage the reverse way round!) as there will then be no current.
  – The **LDR** (Light Dependent Resistor) is a special resistor that will have less resistance in brighter light.
  – The **thermistor** is a special resistor that will have less resistance at higher temperatures.
  – Special **switches** that can be turned on by pressure, being tilted or being next to a magnet.
  – **Capacitors** which are used to store *charge*. **When a current flows into a capacitor the voltage across it increases as charge is stored. If a resistor is connected across a capacitor, current is driven through it and the voltage across the capacitor falls – it is discharged. A larger resistor lets through a smaller current and the fall in voltage takes longer. In a similar way, the time taken to charge the capacitor will be increased if the charging current passes through a larger resistor. Bigger capacitors also take longer to charge and discharge. The capacitor can therefore be used as a simple **timer**.

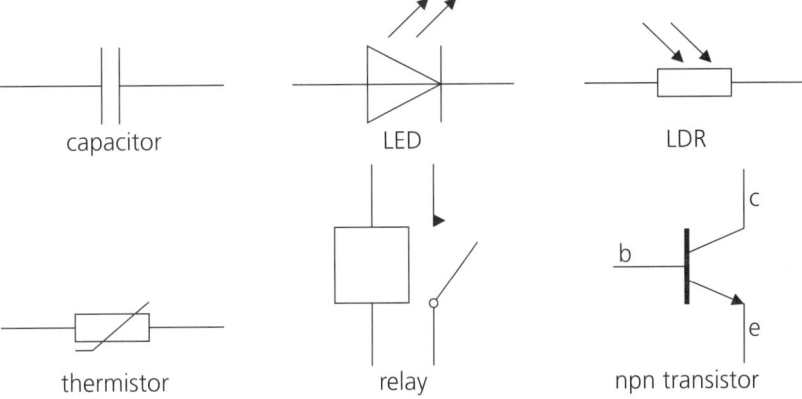

Symbols for the new components

▶ **Transistors** have three connections called the **emitter, collector** and **base**. These will not normally be labelled on the diagram so do remember which is which. The transistor will be switched on by a very small current and then allows a much larger current to flow. It is then a **current amplifier** and can be used as a *buffer* to let low current components like logic gates switch on higher current components like relays.

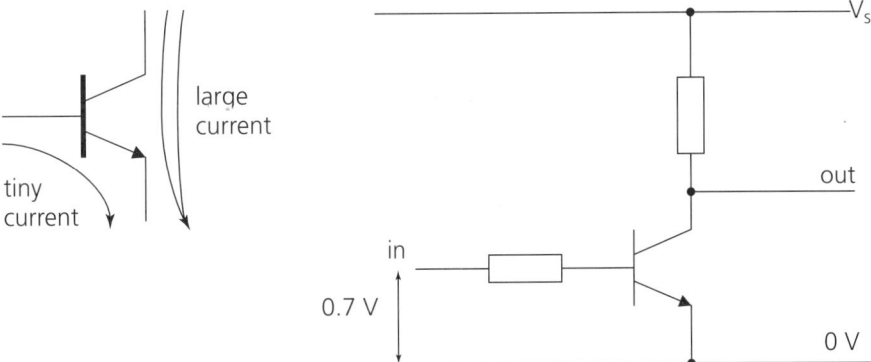

An npn transistor and an amplifier circuit

**In this case a very small current through the base and emitter switches the transistor on and allows a much larger current to flow through the collector and emitter. Since both currents flow through the emitter:

emitter current = collector current + base current.

KEY POINT

$$\text{Current gain} = \frac{\text{collector current}}{\text{base current}}$$

The voltage needed to drive the current though the base-emitter in this way is 0.7 V and the transistor will be on if the voltage is more than this.
**The current gain of the transistor is given by the equation shown.
When the transistor is used as a switch like this inputs are connected to the base and outputs into the collector lead.

▶ **Electronic systems** all have the same layout so that:

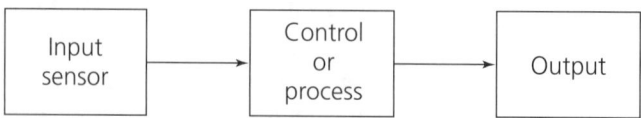

- The **input sensor** can use thermistors, LDRs, switches, microphones or moisture detectors to detect changes in the environment.
- The **control or process** unit can be a transistor or logic gates, and decides what action to take.
- The **output** can be a simple signal from a bulb, buzzer or LED. It could also be a heater or motor or other device controlled by a relay.

▶ **Potential dividers are used to split voltages as shown in the diagram.
Sometimes the two resistors are replaced by a single potential divider with a centre slider so that the output voltage is adjustable.

The potential divider

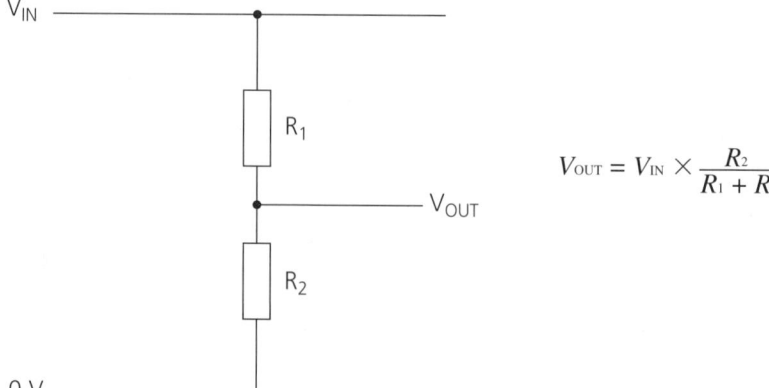

$$V_{OUT} = V_{IN} \times \frac{R_2}{R_1 + R_2}$$

Sometimes you need to know what happens without an accurate calculation. Remember that the output is across R_2 and the bigger the resistor the bigger the share of the voltage that it gets. For example, if R_2 gets larger the voltage across it increases and the output voltage is bigger; but if R_1 gets bigger then it has a bigger share of the voltage and the output voltage gets smaller. Most simple sensors are potential dividers. One half of the divider will be the LDR, thermistor, or other sensor and the other half a variable resistor. The variable resistor lets you set the level of light, temperature etc. that will switch the processor. The circuit shows a light sensor switching a mains light.

A transistor switch

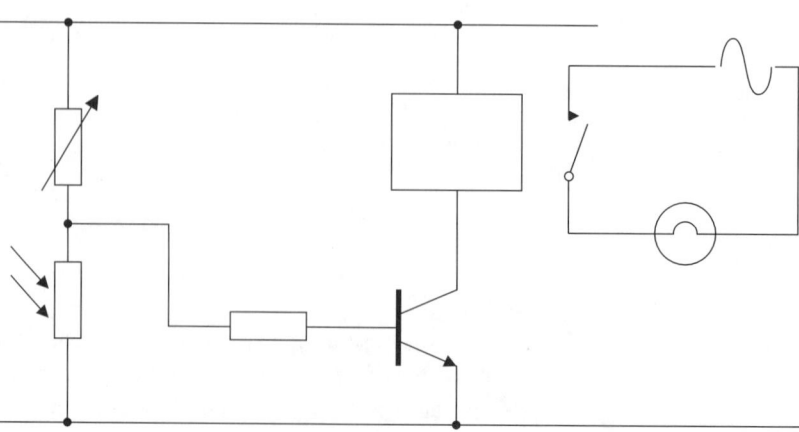

▶ **A logic gate** is an electronic switch and only has its output on (1 or high) if its inputs are in the correct state. This is shown by a truth table for each gate in the following table, together with the symbols. Notice that NAND is the same as AND followed by NOT and that NOR is the same as OR followed by NOT.

Logic gates

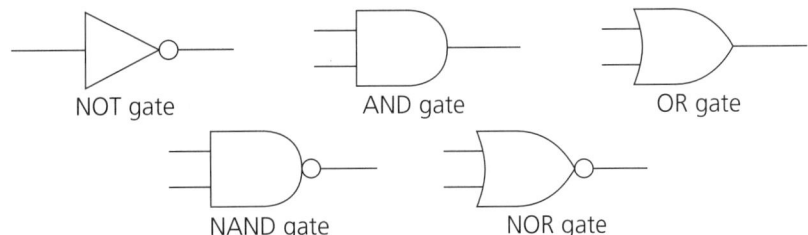

NOT gate   AND gate   OR gate

NAND gate   NOR gate

Truth tables for logic gates

NOT	
Input	Output
0	1
1	0

AND		
Input	Input	Output
0	0	0
0	1	0
1	0	0
1	1	1

OR		
Input	Input	Output
0	0	0
0	1	1
1	0	1
1	1	1

NAND		
Input	Input	Output
0	0	1
0	1	1
1	0	1
1	1	0

NOR		
Input	Input	Output
0	0	1
0	1	0
1	0	0
1	1	0

▶ **A bistable** circuit is one that can stay switched in either of *two* states. A common example is a **latch**.

A latch

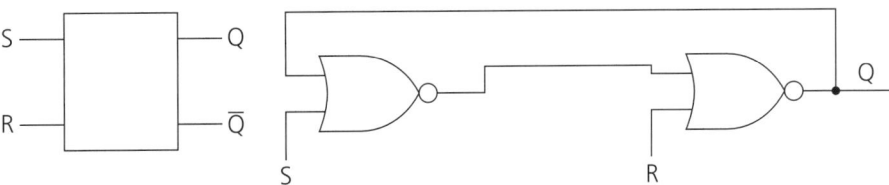

The latch can have its output switched on by making the **set** input high for a short time. The output will then stay high until the **reset** input is made high for a short time, after which the output will stay off until the set input is switched on again. On the diagram, S is the set input, R is the reset input, Q is the output and $\overline{Q}$ is another output that is always opposite to Q. A latch can be made from two NOR gates. A latch might be used to keep a burglar alarm on once triggered. It would stay on until reset by a special key switch or number pad. It can also be used as the memory for one bit of a number, which it 'remembers' until reset. (One bit of a binary number is only ever on or off.)

▶ **Analogue** signals can be continuously changed to any value. Most real, natural things such as temperature and time are like this. Analogue signals can be shown by a pointer that can move to any point along a scale, e.g. the hands on a clock.

▶ **Digital** signals change in steps that can be counted. Most electronic circuits work this way; e.g. a digital watch which counts in seconds. Since you cannot have a reading between the steps, it is important that the steps are small enough to give you the accuracy that you need; e.g. a digital thermometer which goes up in one degree steps may not be sensitive enough.

Useful equations

$$\frac{\text{Voltage across secondary}}{\text{Voltage across primary}} = \frac{\text{turns on secondary}}{\text{turns on primary}} \qquad \frac{V_2}{V_1} = \frac{t_2}{t_1}$$

Emitter current = collector current + base current $I_E = I_C + I_B$

Current gain = $\dfrac{\text{collector current}}{\text{base current}}$ current gain = $\dfrac{I_C}{I_B}$

REVISION ACTIVITY

Make sure that you can do the following short checks and then attempt the main revision questions. Remember to write down any equations that you use.

1   An induced voltage made by moving a wire in a magnetic field can be made larger by _____ or _____.

2   _____ are used to step voltages up or down.

3   An LDR has _____ resistance in brighter light.

4   A thermistor has more resistance when it is _____.

5   A _____ is a component used to store charge.

6   A transistor has three connections called _____, _____ and _____.

7   An electronic system has three parts called _____, _____ and _____.

8   A _____ circuit can stay in either of two states. A common example is the _____.

9   Draw the truth table for an AND gate.

10   A sensor is made from a LDR and a variable resistor as shown.

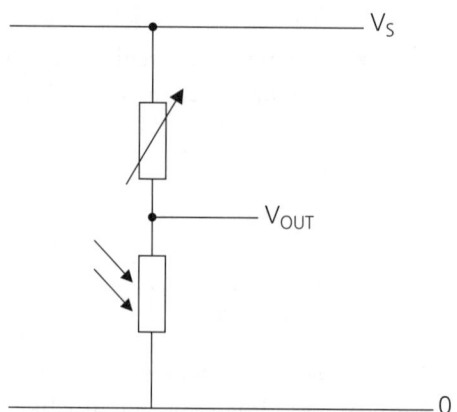

In brighter light the output voltage will get _____ because
_____

EXAMINATION QUESTIONS

Questions that are most likely to be on the higher paper only are marked **

Question 1

(a) The logic gate A in the diagram is an OR gate and is followed by logic gate B which is a NOT gate.

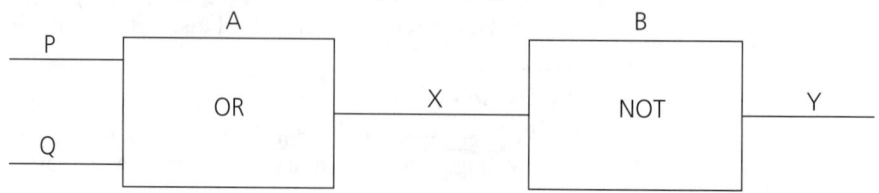

Complete the following truth table to show the outputs X and Y.

Input P	Input Q	Output X	Output Y
0	0		
0	1		
1	0		
1	1		

(b) This combination can be replaced by one gate. State its name.

....

Question 2

(a) The symbol below shows a transistor. Label each of the connections correctly.   [3]

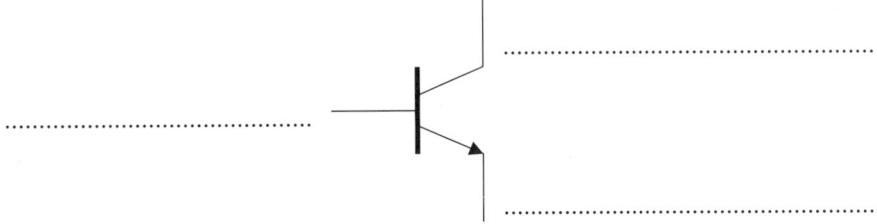

(b) (i) Which one of the following circuits is correctly wired so that the bulb will light?

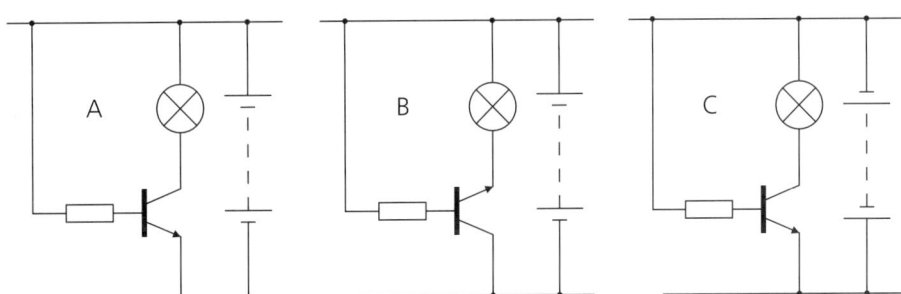

The correctly wired circuit is ....   [1]

(ii) Fill in the following statements to explain why the other two circuits do not light the bulb.

The bulb in circuit .... will not light because

....

....[1]

The bulb in circuit .... will not light because

....

....[1]

**(c) Use the equation $I_E = I_B + I_C$ to find the emitter current I_E in the transistor shown.   [1]

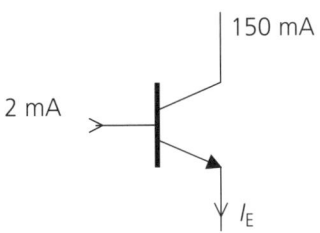

Question 3

The system in the diagram is used on a machine in a workshop to make it safer to operate. The switch is on when the safety guard is properly closed. The light sensor has an output that is high (1) when a hand close to it puts it in shadow.

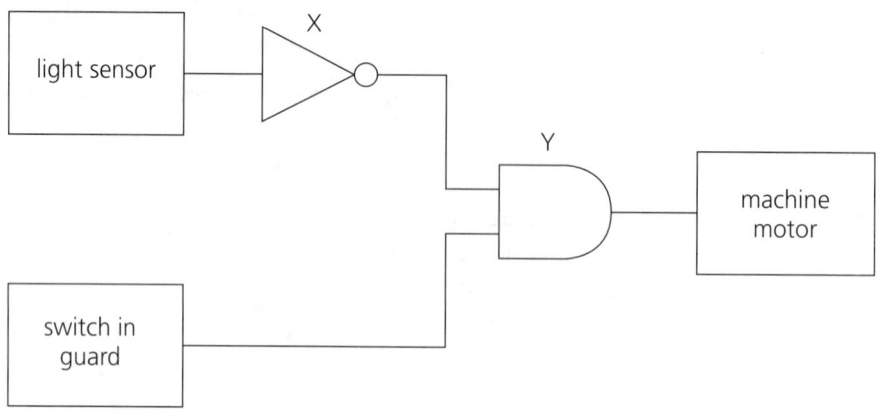

Name logic gate X .... Name logic gate Y .... [2]
Describe carefully the state of the sensors when the motor is running.

....

....

....

.... [2]

Explain what would happen if an OR gate was put into the system instead of gate Y.

....

....

.... [2]

Question 4

A car is fitted with a security device. To start the car, two keys are needed. The first key operates a 'hidden' switch, the second key operates the ignition switch.

If the 'hidden' switch and the ignition switch are both on, the engine will start. Turning the ignition switch without operating the 'hidden' switch will not allow the engine to start and will activate an alarm.

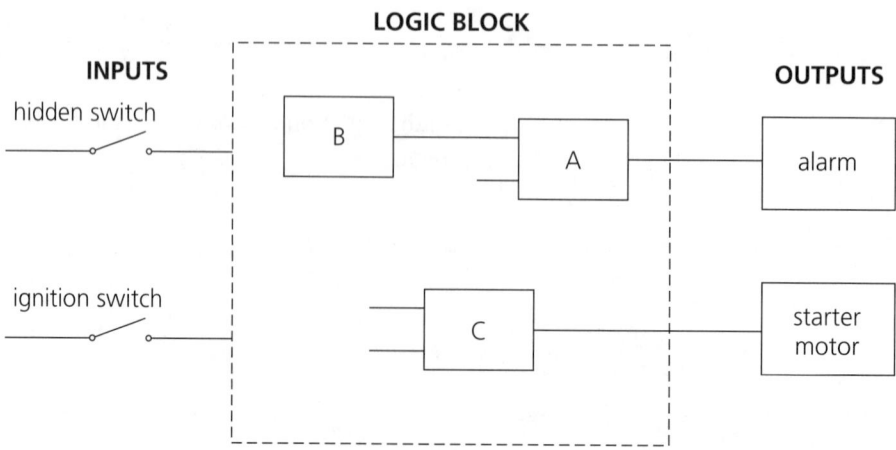

A, B and **C** are the three gates needed for the design of the logic block.

(a) Identify gate

    (i) **A** ....   [1]

    (ii) **B** ....   [1]

    (iii) **C** ....   [1]

(b) **Show on the diagram** how the **INPUTS** and **LOGIC GATES** are connected together.   [2]

(c) The logic gates used in the logic block are called digital devices. Explain what this means.   [1]

....

....

WJEC 1996

3 Forces and motion

Topics marked ** are usually on the higher paper and may not be on all syllabuses. Quite a lot of this work has equations so that you can do simple calculations. These have also been collected together at the end of the notes so that they are easier to find and check. Do make sure that you learn all the equations that you will need and that you are confident about using them.

TOPIC OUTLINE

Motion

▶ The motion of an object can be shown by a **distance–time graph**. You should recognize the following shapes.

Distance–time graphs

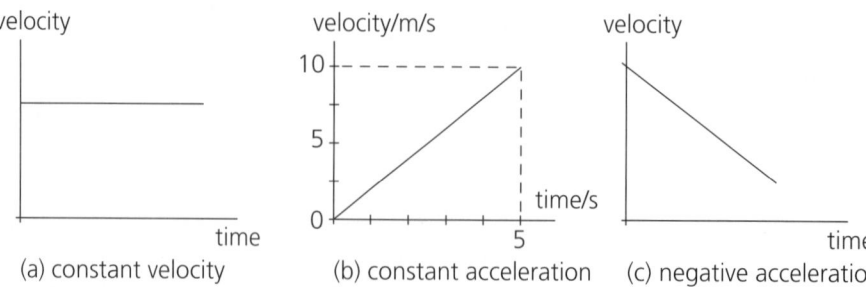

(a) stationary (b) constant speed (c) accelerating

KEY POINT

$Speed = \dfrac{distance}{time}$

Steeper graphs show faster speeds.

▶ You can work out **speed**s by using the speed equation on the numbers from the axis of the graph. In (b) the speed will be $s = d/t = 20/5 = 4$ m/s. This is called finding the *gradient* of the graph.

If the graph slopes downwards the speed is negative which means that the object is going backwards.

▶ **Velocity** is the speed in a particular *direction*.

▶ When velocity changes there is an **acceleration**.

KEY POINT

Acceleration =

$\dfrac{change\ in\ speed}{time\ taken}$

You can work out the **acceleration** of an object from a **velocity–time graph**. Steeper graphs mean faster accelerations.

Velocity–time graphs

(a) constant velocity (b) constant acceleration (c) negative acceleration

▶ **You can work out the acceleration from the numbers on the axes of a velocity–time graph. In (b) the acceleration will be a = change in velocity/time taken = $^{10}/_5 = 2$ m/s²
This is finding the gradient of the graph again.
**Negative accelerations mean that the object is slowing down and the graph will be sloping from top left towards bottom right.

▶ **You can also find the distance travelled by finding the area under the graph.

Forces

► Forces can pull, push, twist and deform things.
Forces can also change velocity by changing the speed or the direction of an object. The object has then been accelerated.

► Objects will not change their velocity unless there is a force to make it happen. This is called **Newton's first law**. If you are travelling on a bus which suddenly stops you will keep on going forward until there is a force to stop you. Note that you are *not* being thrown forward – the problem is that you aren't being stopped!

► When forces on an object are **balanced** (equal and opposite) it will either remain stationary or keep going at the same velocity. Sometimes an object will accelerate until the *driving forces* (such as weight or a pushing force) are balanced by the *counter forces* (such as friction or air resistance) and it then reaches a **terminal velocity**. This will happen to a free-fall parachutist.

► Larger forces will be needed to produce bigger accelerations or accelerate bigger masses. All forces are measured in *newtons* (N).
The equation in the box is one version of **Newton's second law.**
A force of 1 **newton** will accelerate a mass of 1 kilogram at $1m/s^2$.

► It is important to realize that forces in accidents and collisions can be changed if the time taken for the collision is changed. Remember that the force is proportional to the acceleration. If you can make a traffic accident collision take longer then the acceleration is smaller and the forces are also made smaller. *Seat belts, crumple zones, air bags* all make the force spread over a longer time so that the damage to the body is made smaller. Egg boxes and cycle helmets will also crumple to make the time longer and the damage to the contents less.

► **Newton's third law** says that when two objects collide the forces that they put on each other will be equal and opposite. The same idea applies when things explode or in a rocket or a jet engine.

► **Friction** is caused by an object moving through a fluid such as air or water when it will increase with speed. It is also caused by two solids sliding over each other and the friction force then depends on the force between the two surfaces and the type of surface – NOT on the area. The friction force is always in the opposite direction to the movement or the force being applied, it is a **counter force**. The friction force will also depend on the speed of the object. (So the free-fall parachutist can have an air resistance to balance his speed if he is falling fast enough!)

► **Turning forces** have many important practical uses. The size of a turning force is called its **moment**. The moment depends on the size of the force and its distance from the pivot. If the force goes through the pivot the moment will be zero.

► When an object is balanced (in equilibrium) the total moments about any point is zero. We can say that **total clockwise moments = total anticlockwise moments**. This is called the **Principle of Moments**.

► **Levers** use the same principle. The *effort* force is usually at a greater distance from the pivot than the *load*. The larger load can then be moved by a smaller effort, e.g. a crowbar or a wheelbarrow. These are examples of machines which are **force multipliers**. In a few cases the load is further from the pivot than the effort and is therefore smaller than the effort. Such a machine is a **distance multiplier.**

► The **centre of mass** of an object is the point at which we can assume all the mass of an object acts. We can also assume that all the weight will act at this same place and it is then called the *centre of gravity*.
If you suspend a thin sheet so that it can turn freely then it will hang so that its centre of mass is directly below the suspension point. If you draw a vertical line from the suspension point the centre of mass must lie somewhere along it.

Suspend the sheet from another place and repeat the process. The centre of mass will be where the lines cross. A third line makes a good check.

▶ **Pressure** is the force acting on each unit of area.
   A pressure of 1 newton on 1 m² is called 1 pascal (Pa).
   The force may be caused by weight acting on the area or by the collisions of gas molecules on the surface. As you go deeper down into a liquid the pressure increases. As you go further up in the *atmosphere* the pressure gets less because there are less air molecules hitting the surface.

▶ Liquids can't be squashed into a smaller space so a pressure acting on any part of a liquid is felt in all directions all the way though the liquid. This is used in transmitting forces from one place to another (e.g. car brakes) and to get large forces to act in a hydraulic press.

A hydraulic press

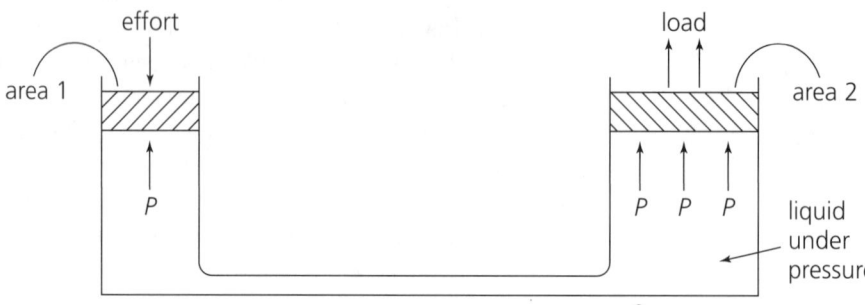

In the hydraulic press in the diagram load = effort × $\dfrac{\text{area 2}}{\text{area 1}}$

The force on the load piston of the press is equal to the force on the effort piston multiplied by a number equal to the area of the load piston divided by the area of the effort piston. This can be a large number but the greater force is moved through a corresponding smaller distance.

▶ **Hooke's law** says that a spring or a wire will stretch by a length called the extension which is proportional to the load force applied. Provided that the force does not get larger than the *elastic limit*, the spring or wire will return to the original size and shape when the load is removed. Bending beams also obey the same sort of rule.

A graph showing Hooke's law

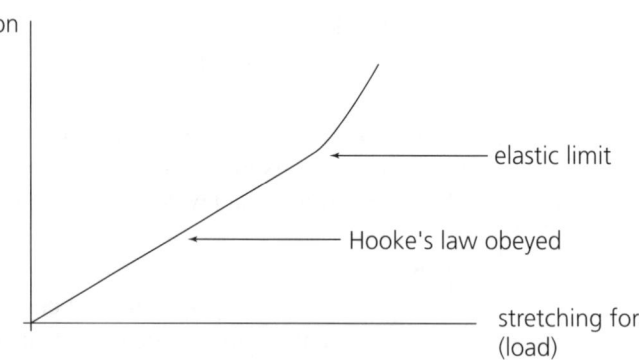

Hooke's law is often tested by stretching rubber bands or springs by hanging weights from them and measuring their extension.

▶ ** **Momentum** depends on mass and velocity. The greater the force the faster it will change momentum.

▶ ** When objects collide the total momentum before the collision is the same as the total momentum after the collision. This is called **conservation of momentum**. The two objects exert forces on each other that are equal in size but opposite in direction. This will mean that each of the two objects has a change in momentum which is equal in size but opposite in direction. This is an alternative way of looking at Newton's third law. In an elastic collision the total kinetic energy would also remain the same (such as a collision between two molecules). Most collisions are not perfectly elastic and the less elastic the collision the more kinetic energy is lost.

▶ ****Circular motion** An object that is moving in a circle is changing its velocity all the time. The speed may be constant but the direction is changing. A force *towards the centre* called a **centripetal force** is needed. The centripetal force depends directly on the mass and speed of the object and inversely on the radius of the circle.

This centripetal force is provided by gravity for planets and satellites and by electrostatic forces between electrons and the nucleus in atoms.

▶ **Boyle's law** is about the volume and pressure of a gas when its temperature stays the same. It says that the volume is inversely proportional to the pressure. Another way to say this is that the pressure × volume will always give the same answer.

Remember that the law only works at the same temperature and when the mass of gas stays the same. The reason is that the pressure is caused by the molecules of gas hitting the walls. When the volume is made smaller the molecules hit the walls more often and the pressure rises in proportion.

The law can be shown by measuring the pressure and volume of gas in a syringe or by a special apparatus which uses oil to trap, and put pressure on, some dry air. The graph of pressure against volume will be a curve. Plotting pressure against 1/volume will give a straight line graph. Sometimes gas pressure is measured in atmospheres instead of pascals. One atmosphere is the average value of air pressure at sea level.

▶ ****The gas equation** can be used to find out what happens to a gas when volume, temperature or pressure change.

It is important that the temperatures are in kelvin (K). To convert from °C to kelvin, add 273, e.g. 27 °C = 300 K.

KEY POINT

Pressure × volume =
A constant or
$P_1V_1 = P_2V_2$

KEY POINT

$\dfrac{pressure\ 1 \times volume\ 1}{temperature\ 1} =$

$\dfrac{pressure\ 2 \times volume\ 2}{temperature\ 2}$

$\dfrac{P_1V_1}{T_1} = \dfrac{P_2V_2}{T_2}$

An example

100 cm³ of a gas were collected in a science experiment. The gas was under a pressure of 3 atmospheres and at a temperature of 127 °C. How much would have been collected at 1 atmosphere pressure and 27 C°?

$$\frac{P_1\,V_1}{T_1} = \frac{P_2\,V_2}{T_2}$$

$$\frac{1 \times V_1}{300} = \frac{3 \times 100}{400}$$

$$V_1 = \frac{3 \times 100 \times 300}{400}$$

$$V_1 = 225\ cm^3$$

The volume would have been 225 cm³

**Equations of motion

These laws apply when an object has constant acceleration. They are not on all syllabuses. They are:

$$v = u + at \qquad\qquad v^2 = u^2 + 2at \qquad\qquad s = ut + \tfrac{1}{2}at^2$$

where v = final velocity, u = starting velocity, t = time, s = distance, a = acceleration.

Useful equations

Speed = $\dfrac{\text{distance}}{\text{time}}$	$s = \dfrac{d}{t}$	speed units are m/s
Acceleration = $\dfrac{\text{change in speed}}{\text{time taken}}$	$a = \dfrac{v}{t}$	acceleration units are m/s²
Force = mass × acceleration	$F = ma$	Force units are N
Moment = force × distance to pivot	$m = Fd$	Moment units are Nm

Useful equations

Pressure = $\frac{\text{force}}{\text{area}}$ $\qquad P = \frac{F}{A}$ Pressure units are N/m² or Pa

****Momentum** = mass × velocity $\quad M = mv$ Momentum units are kgm/s

Pressure × **volume** = constant $\qquad P_1V_1 = P_2V_2$ *only* for gas *at const. temp.*

**$\frac{\text{Pressure 1} \times \text{volume 1}}{\text{temperature 1}} = \frac{\text{pressure 2} \times \text{volume 2}}{\text{temperature 2}} \qquad \frac{P_1V_1}{T_1} = \frac{P_2V_2}{T_2}$

****Force** = $\frac{\text{change in momentum}}{\text{time taken}} \qquad F = \frac{\Delta M}{\Delta t}$

****Centripetal force** = $\frac{mv^2}{r}$

REVISION ACTIVITY

Make sure that you can do the following short checks and then attempt the main revision questions. Remember to write down any equations that you use.

1 A tennis ball travels 25 m in 2 s. What is its speed?
2 A car speeds up from 10 m/s to 20 m/s in 3 s. What is its acceleration?
3 Explain how a free-fall parachutist can reach a constant speed.
4 How does the force on an object depend on the mass of the object and the acceleration produced?
5 Name one way of making the force on the passengers in a car accident smaller. How does this safety process work?
6 What is meant by centre of mass? Where would the centre of mass of a bicycle wheel be?
7 What is meant by *elastic limit*?
8 What must be kept the same for Boyle's law to be true for a gas?
9 Write down the equation that links pressure, force and area.
10 What is different about the way in which a pressure affects a liquid and a gas?
11 ** Write down the equation that links momentum, mass and velocity.
    What is the momentum of a car moving at 10 m/s if its mass is 1200 kg?

EXAMINATION QUESTIONS

Questions that are most likely to be on the higher paper only are marked **

Question 1
An object increases its speed uniformly from **rest**.
After 4 seconds its speed is 20 m/s.
Find its rate of change of speed (acceleration).
Show clearly how you get your answer.

....

....

....

....

....

Rate of change of speed (acceleration) = _____ [3]

CCEA 1996

Question 2

(a) A glass window pane covers an area of 0.6m². If the pressure of the atmosphere is 100 000 N/m² calculate the force exerted by air pressure on the outside of the window pane. Write down the formula that you use and show your working. [3]

....

....

....

....

....

(b) Explain why the window does not break under this force. [1]

....

....

WJEC 1996

Question 3

X

weight

A sky diver jumps from an aeroplane and free falls without opening the parachute. The acceleration at the start of the fall was 10 m/s². The sky diver was acted upon by two forces as shown. Force **X** is extremely small at the start but increases during free fall.

(a) (i) Name the force **X**. [1]

....

(ii) State what effect the increasing force **X** has on the acceleration of the sky diver. [1]

....

(iii) Explain why the sky diver eventually stops accelerating and moves at a steady speed. [1]

....

....

(b) After opening the parachute, the sky diver hits the ground at 4 m/s and he comes to rest 0.8 s after making contact with the ground. Writing down the formulae that you use and showing your working, calculate:

(i) the deceleration at the ground [3]

....

....

....

(ii) the decelerating force, if the mass of the sky diver is 60 kg. [3]

....

....

....

WJEC 1996

Question 4

(a) When car drivers see an emergency in front of them, there is always a short reaction time before they apply the brakes. The distance travelled during this delay is called the 'thinking distance'.

  (i) A driver is travelling at 20 m/s on a dry road. He takes 0.6 s to react. Calculate his 'thinking distance'.

  ....

  .... [2]

  (ii) At this speed the braking distance of the car is 30 m. What is the TOTAL stopping distance for the car?

  .... [1]

  (iii) Driving conditions change. The roads may be wet. The driver may be tired. The driving conditions may change the thinking and braking distances. Complete the table, stating whether the distances will increase, decrease or remain the same.

Condition	Thinking distance	Braking distance
Wet roads		
Tired driver		

[4]

  (iv) Add to the diagram below the direction of the braking force acting on the car when the brakes are applied.

Direction of motion

(b) The diagram below shows the hydraulic braking system of a car. When the driver's foot pushes on the brake pedal, a large force is produced by the brake pads on the discs and the car stops.

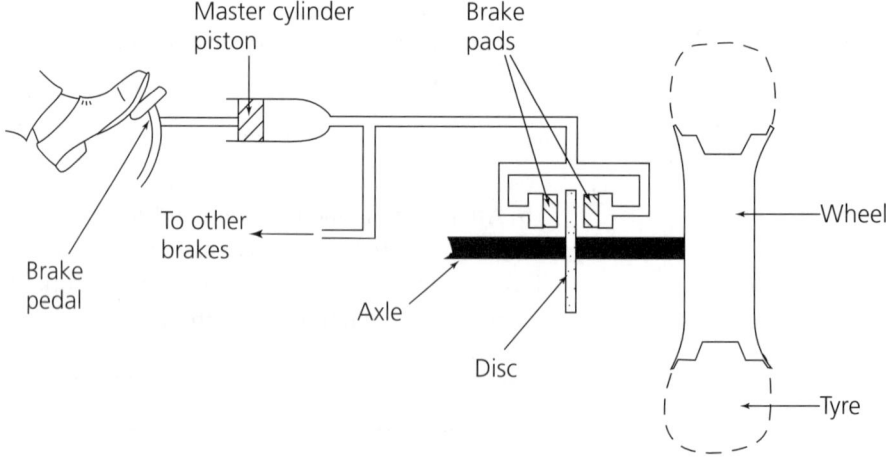

  (i) Explain how a force is produced by the brake pads when the brake pedal is pushed.

  ....

  ....

  .... [3]

(ii) It is normal to use brake fluid (an oil) in the pipes to the brake pads. Explain why air would not be suitable.

.... [1]

(iii) The area of the master cylinder piston is 3.0 cm². The foot pushes with a force of 120 N. Calculate the pressure in the master cylinder.

....

.... [2]

(iv) State the pressure exerted by the brake fluid on the brake pads.

.... [1]

(v) The area of the brake pads is 50 cm². Calculate the total force exerted on the disc when the foot pedal is pushed.

....

.... [2]

(vi) Describe the energy transfers taking place when the car is stopped by the brakes.

....

....

....

.... [2]

London 1996

Waves, light and sound

Topics marked ** are usually on the higher paper and may not be on all syllabuses.

TOPIC OUTLINE

Remember that both light and sound are waves and will have the properties of waves. Many properties, such as reflection, are the same for both.

▶ **Transverse waves** have their vibration at right angles to the direction of travel. All electromagnetic waves are transverse (including **light**) and so are ripples on water.

▶ **Longitudinal** waves have their vibration with and against the direction of travel. This produces compressions that travel along. This is the way that air vibrates as a **sound** wave passes.

▶ **Waves** will carry **energy** outward from the source but the particles themselves only vibrate and do not move along with the wave. Most waves, including sound waves, begin at a source that *vibrates*.

▶ There are several measurements that can be made on a particular wave.

– The distance between one point on a wave and the next point where the vibration is the same is called a **wavelength** (λ). It will be measured in ordinary length units (m or mm).

– The largest distance moved by a particle in the wave is the **amplitude** (a). A bigger amplitude will mean that the wave carries more energy.

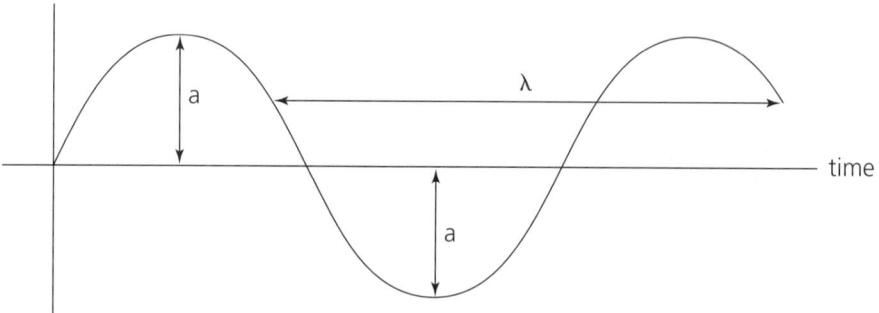

– The number of waves each second that pass a particular place is called the **frequency** and in measured in hertz (Hz). A greater frequency will also mean that the wave carries more energy. This is especially important in electromagnetic waves where the short wavelengths such as X-rays or gamma radiation carry most energy.

– The **speed** of a wave can always be found by using the equation shown.

KEY POINT

wave speed =
   frequency $\times$ wavelength
$v = f\lambda$

An example

A water ripple has a frequency of 10 Hz and a wavelength of 2 cm. How fast is it travelling?

wave speed = frequency $\times$ wavelength
= 10×2
= 20 cm/s

Light

▶ **Light waves** are transverse electromagnetic waves and they can travel through
a vacuum. Their speed in a vacuum is 300 000 000 m/s.
▶ Waves can be **reflected**. Waves on ropes, springs and across water are reflected.
Reflected sound waves are heard as **echoes**.
  – Light will be reflected from a flat shiny surface (a plane mirror) so that the
  angle of incidence and angle of reflection are the same.
  – Light being reflected like this will produce an **image** the *same size* as the
  object but turned sideways (*laterally inverted*) and the *same distance* as the
  object from the mirror but *on the other side*. The light *appears* to come from
  an image like this but it does not actually pass through it. It is called a
  virtual image and cannot be found on a screen.

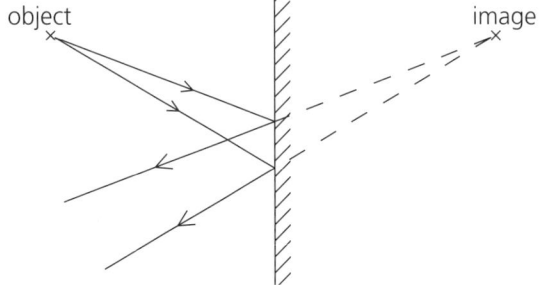

▶ Waves are **refracted** (change direction) when they *change speed* as they go
from one material into another (unless they meet the boundary at 90°). They will
bend towards the normal as they are slowed down and away from the normal
when speeded up. Water waves go slower in a shallower depth of water. Light
waves go slower in more dense materials.
  – Sound can also be refracted (e.g. by balloons of gas) – it will be speeded up
  by more dense materials.
  – When a ray of light travels from glass or perspex or water into air it will be
  refracted *away* from the normal so that the angle of refraction is greater than
  the angle of incidence. When the **critical angle** inside the material is reached
  the angle of refraction is 90° and cannot become larger. If the angle of
  incidence is increased the ray of light will be **totally internally reflected**.

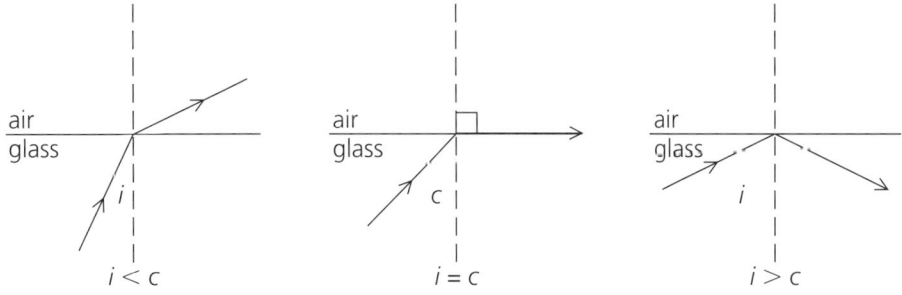

  – The critical angle for glass or perspex is about 42° so an internal ray at 45°
  will be internally reflected. This is used in optical fibres and also in some
  prisms that change the direction of light beams.

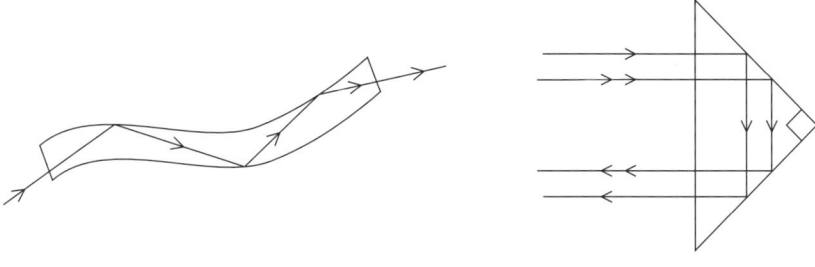

(a) optical fibre   (b) total internal reflection in a prism

Sound

- ▶ **Sound waves** are longitudinal and must have a material to travel in so that the compressions can be made. Their speed in air is about 330 m/s and about four times as fast as this in water.
- ▶ Sounds are always produced by a **vibrating source**. A normal ear can detect sounds from about 20 Hz up to 17 kHz.
  - – Sound waves with *larger amplitudes* are *louder*.
  - – Sound waves with a *higher frequency* will have a *higher pitch*.
  - – Frequencies and amplitudes can be shown by picking up the sound with a microphone and showing it on an oscilloscope.
  - – **Noise** is unwanted sound. The intensity of noise and other sound is often measured in **decibels (dB)**. The quietest whisper that you can hear would be 0 dB and 120 dB will start to hurt your ears.
- ▶ When waves travel through a narrow gap they are spread outwards. This is called **diffraction**. The diffraction will be greater if the gap is smaller when compared to the wavelength.

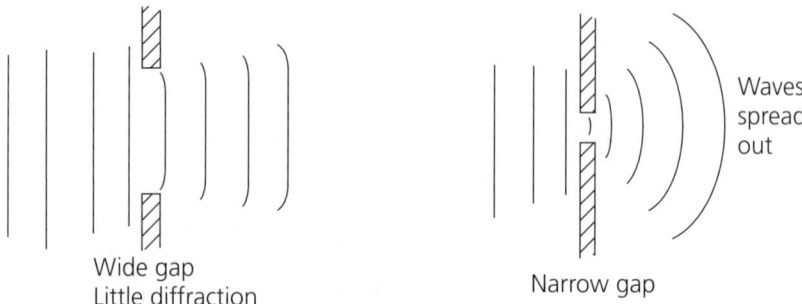

Wide gap
Little diffraction

Narrow gap

Waves spread out

Red light is diffracted more than blue through the same gap because it has a longer wavelength. Diffraction is the reason why radio signals can sometimes be heard behind hills.

- ▶ Sounds with a higher frequency than we can hear are called **ultrasound.** Ultrasound is used to produce reflection pictures for 'scans' of pregnant women where X-rays would be dangerous. Ultrasound is used because it has a shorter wavelength and is therefore diffracted less than ordinary sound. In a similar way ultrasound is used to detect flaws in metal castings. Ultrasound can also be used for industrial cleaning of delicate mechanisms.

Lenses

- ▶ **Lenses** are used to make rays of light go in the directions that we want. A lens that has surfaces which are more curved will bend the light more.
  - – A **concave** (diverging) lens will refract parallel light so that it spreads apart as though it had come from a *focus*. The image from a concave lens will always be smaller than the object, the correct way up. This sort of lens is used to correct short sight.
  - – A **convex** (converging) lens will refract parallel light together so that it passes through a *focus*. It will be thicker in the centre than at the edges.
- ▶ **The image formed by a convex lens** depends on its distance from the object.
  - – If the object is *nearer than the focus* the image will be magnified, the correct way up and *virtual*. Use: the magnifying glass.
  - – If the object is slightly *further away than the focus* the image will be magnified, inverted and real (can be found on a screen). Use: the **projector**.
  - – When the object is at *two focal lengths* from the lens, its image will be the same size, real but inverted.
  - – As the object distance increases *beyond two focal lengths* the image distance decreases and the image is smaller than the object, inverted and real. Uses: **camera** lens, **eye** lens. (See chapter on electromagnetic spectrum.)

▶ The distance from a lens to its focus is called the **focal length**. An approximate focal length for a converging lens can be found if you use it to focus a distant object on to a white screen or wall. Measure from the lens to the image, which should be clear and sharp. The more distant the object the better the answer.

Useful equations

Wave speed = frequency × wavelength $v = f\lambda$

★ REVISION ACTIVITY

Make sure that you can do the following short checks and then attempt the main revision questions. Remember to write down any equations that you use.

1 Noise is _____.

2 Waves carry _____ outwards but the particles only _____ and do not move along with the wave.

3 Sound always starts at something that is _____. It cannot travel through a _____.

4 _____ is caused by a change in the speed of a wave.

5 A wave will be totally internally reflected if it is in the _____ dense material and the angle of incidence is _____ than the _____ angle.

6 A wave has a wavelength of 0.5 m and a frequency of 600 Hz. What is its speed?

7 Blue light will be diffracted _____ than red light at the same opening because it has a _____ wavelength.

8 The image from a _____ lens will *always* be the correct way up and _____ than the object.

9 All waves reflect so that the angle of incidence = _____.

10 A lens that is thicker in the middle will be a _____ lens and will make parallel light converge to a _____.

11 The distance from the centre of a lens to its focus is called the _____ _____.

12 Explain briefly how to find the approximate focal length of a convex lens. How will you make the measurement as exact as possible?

13 Name a use for optical fibre and name two advantages of its use in this application.

14 A lens used as a magnifying glass will be a _____ lens and the object will be _____ to the lens than the focus. The image produced will be magnified, _____ and erect.

? EXAMINATION QUESTIONS

Questions that are most likely to be on the higher paper only are marked **

Question 1

In a thunderstorm the thunder and the lightning are made at the same time. Which of the following statements explains why we usually see the lightning flash before we hear the thunder?

A Light travels much faster than sound.   B Light travels in straight lines.

C Sound and light are waves and both travel at the same speed.

D Sound does not travel well in air.

E The rain in the storm slows the sound down.

Question 2

Which of the following statements is **not** correct?

A All sounds start at something which vibrates.

B A sound with a higher frequency will sound louder.

C Sound cannot travel through a vacuum.

D The loudness of a sound depends on the amplitude of the sound wave.

E Sound waves are longitudinal.

Question 3

Complete the following diagram to show the path of the ray of light after it
strikes the mirror.   [1]

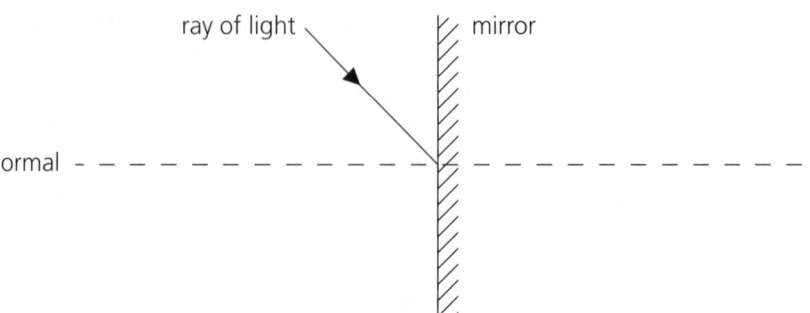

Question 4

(a) Complete the following diagram to show the path of the ray of light after it
    enters the water.   [1]

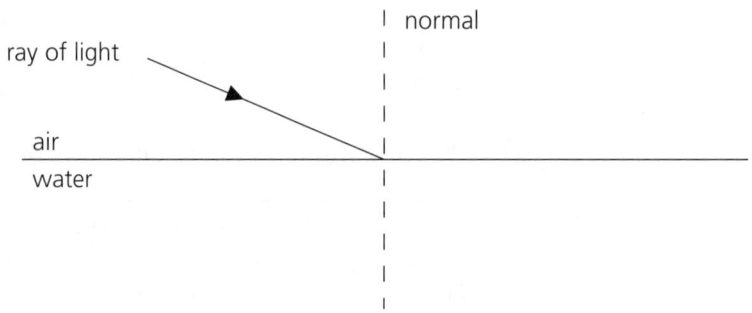

(b) Explain why the ray of light is refracted as it enters the water.   [2]

....

....

....

Question 5

An image is formed on a white screen by a convex (converging) lens. State **two**
things that will happen if the object is moved towards the focus of the lens.

1 .... [1]

2 .... [1]

Question 6

(a) A signal generator connected to a loudspeaker produces a sound wave. With
    the frequency of the signal generator set to 2000 Hz the sound wave has a
    wavelength of 0.17 m in air.
    Calculate the speed of sound in air.

    ....

    ....

                                              Speed = ...m/s [3]

(b) The speed of sound in water is 1400 m/s.
A sound wave has a frequency of 2000 Hz.
Calculate its wavelength in water.

....

....

....

Wavelength = ...m/s [3]

(c) Echo sounders are used at sea to locate underwater objects, such as submarines. The diagram below shows how the echo sounder works.

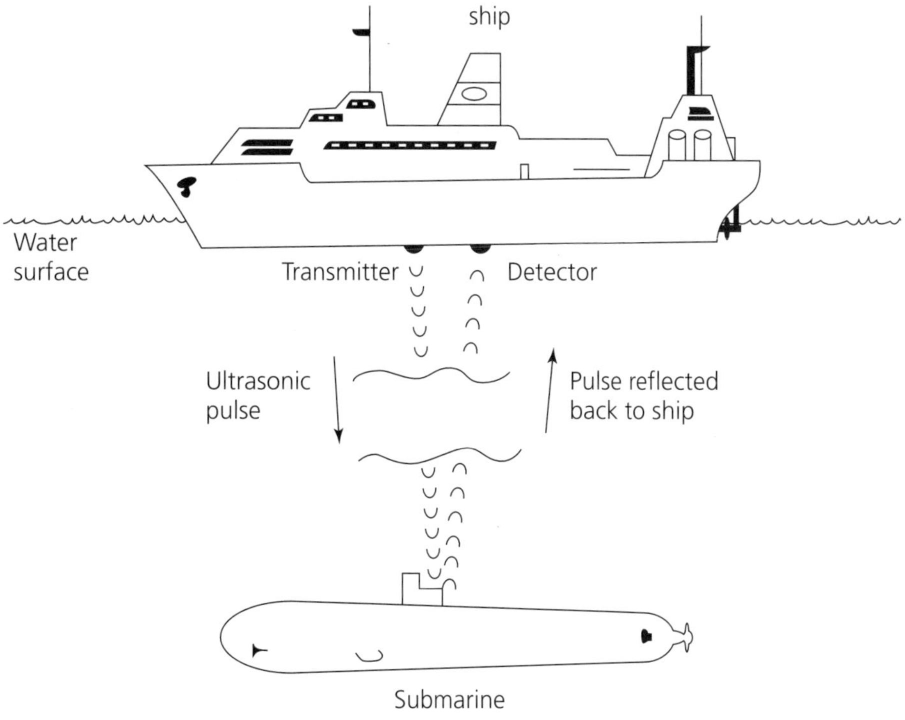

(i) What are ultrasonic waves?

....

....

(ii) The pulse travels from the transmitter to the submarine and back to the detector. The time taken is 0.1 s.
Calculate the distance between the submarine and the ship.

....

....

....

....

Distance = ...m

(iii) State **one** other use for ultrasonic waves.

....

.... [7]

NEAB 1996

Question 7

(a) A source which produces *ultrasonic* waves is placed in front of a large metal sheet with a narrow gap in it. The waves are detected by a microphone on the other side of the metal sheet. The waves detected are then shown on the screen of an oscilloscope.

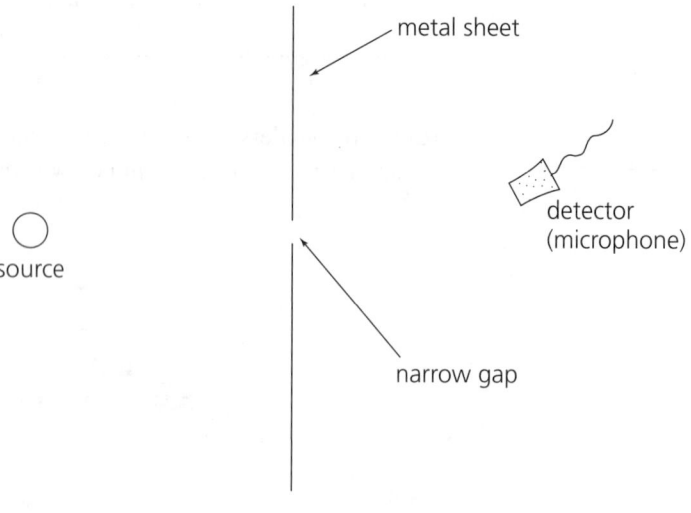

Like sound waves these waves are *longitudinal*.
What is meant by longitudinal?

....

.... [1]

(b) How would you use the apparatus in the diagram to show that the waves are *diffracted*.

....

....

.... [3]

(c) If the frequency of the waves from the source is increased how would you expect the results to change?

....

.... [1]

5 The Earth and space

Topics marked ** are usually on the higher paper and may not be on all syllabuses.

The Earth

▶ **The structure** is known by studying the paths of seismic vibrations (earthquake waves).

The structure of the Earth

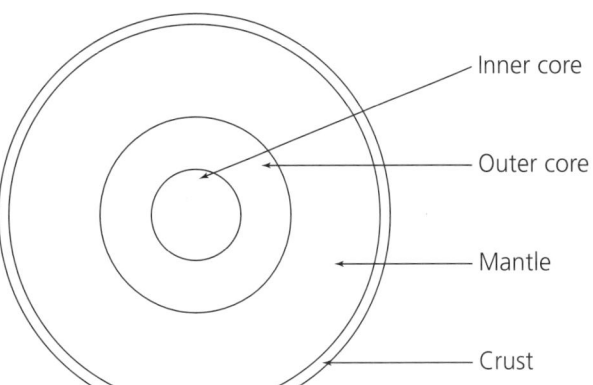

- Inner core
- Outer core
- Mantle
- Crust

  – The **crust** is solid but is relatively thin.
  – The **mantle** is very viscous and its density increases with depth. It goes down to almost half the Earth's diameter.
  – There is a **central core**. The outer part of this core is liquid and the central part is solid.

▶ The Earth spins on its axis every 24 hours. This decides the length of one **day**. The side of the Earth facing the Sun is in daylight and the other side is in night.

▶ The Earth orbits the Sun once each **year**.

▶ The Earth is tilted on its axis so that either the top or bottom half gets more energy from the Sun and has **summer** while the other half is getting less energy and is having **winter**.

Tilt and seasons

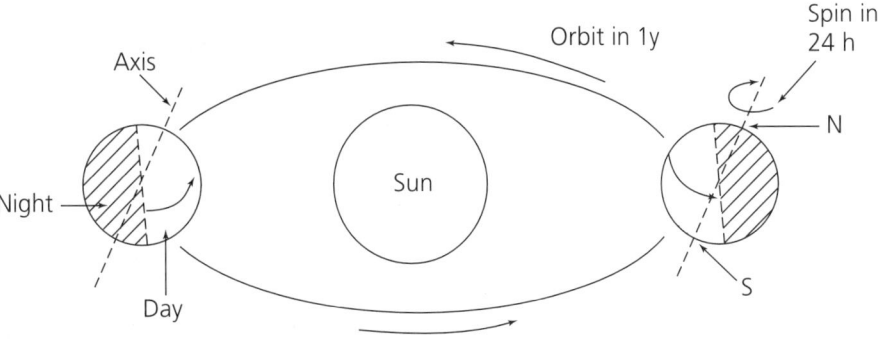

► **Tides** in the oceans on the surface of the Earth are caused by the pull of the Moon's gravitational force on the water. This makes the water bulge slightly on each side of the Earth. As the Earth rotates once each day the water at a particular place will rise and fall twice.

Tides

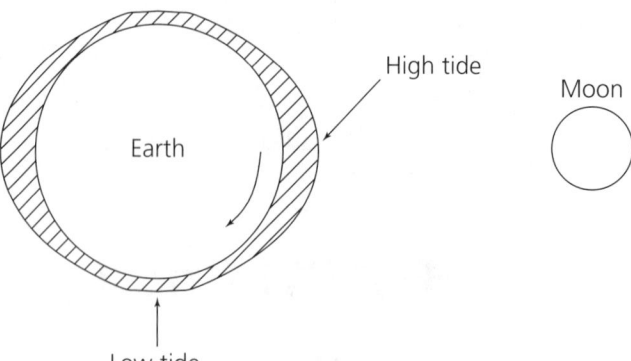

There will be a bigger tide when the Moon and the Sun are in line and both pull in the same direction. This is called a **spring tide** (remember that this has nothing at all to do with the seasons). When the directions of the Moon and the Sun are at 90° the gravity force will be at its weakest and the smallest tides, called **neap tides**, will be produced.

Neap and Spring tides

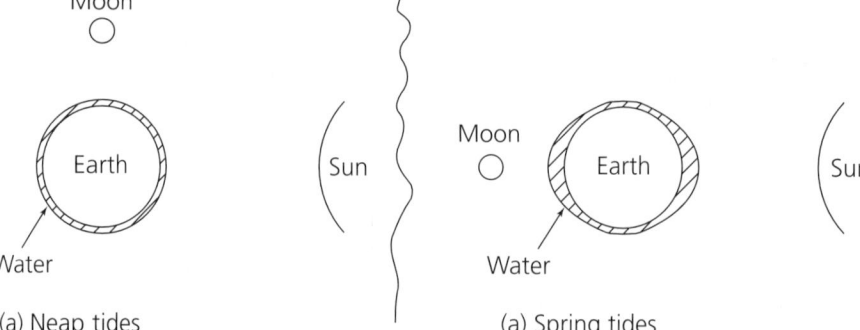

(a) Neap tides   (a) Spring tides

Seismic waves

If you are also studying Chemistry you should read this section together with the work about the Earth and its rocks.

► **Earthquakes** travel through the Earth as three types of wave. They are detected by instruments called *seismographs*. The place where the earthquake occurs is called the **focus** and the point directly above this on the Earth's surface is called the **epicentre**.

  – **P waves** are fastest and are longitudinal. These waves can travel through any part of the Earth but may bend by refraction as they go through places of different density in the mantle.

  – **S waves** are transverse and only travel through solids. They cannot travel through the liquid of the outer core – the shadow that it leaves allows scientists to work out its size.

  – **L waves** are the waves that travel round the crust. They arrive last but do most damage to buildings and other structures. Their greatest effects are local – i.e. close to the epicentre.

  – **The P and S waves travel faster though more dense material. Their path will therefore curve as they go down into the mantle and it becomes more dense. (See the notes on refraction in the section about Waves.) When they go from solid to liquid the P waves will change direction significantly. This happens again when the waves meet the solid core. Scientists used the change in direction to work out that the liquid outer core has a diameter of just over half that of the Earth.

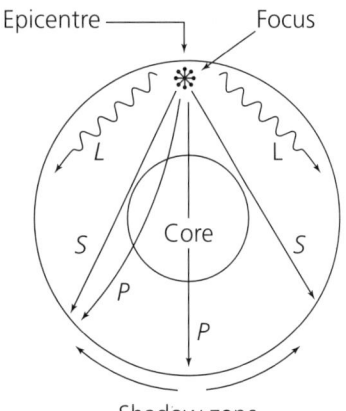

Epicentre — Focus

Seismic waves

Core

Shadow zone

Space

Satellites

▶ A *planet* orbiting its star, a *moon* orbiting its planet or a *man-made satellite* orbiting the Earth are all smaller objects orbiting a larger one. They take longer to go round if the orbit is larger. Each satellite is attracted towards the larger object in the centre by gravity and this provides the **centripetal force** (force *towards the centre*) that is needed to keep the satellite in orbit.

▶ An **orbit** is the path that is taken by a satellite as it goes round the larger object. For man-made satellites going round the Earth the path can be circular. For the planets round the Sun the orbit is an ellipse.

▶ A **polar satellite** goes round the Earth in an orbit over the poles. As the Earth spins underneath the satellite passes over all of the Earth after a few orbits. They can therefore monitor all that is happening on the Earth's surface including keeping a check on crops and military movements.

▶ A **geostationary satellite** orbits the Earth a long way above the equator. It orbits once every 24 hours, so it stays above the same place on the Earth's surface as that also rotates at the same rate. These satellites are especially useful as communications satellites for TV and telephones.

Solar systems facts

▶ **Planets** move in orbits round the Sun. They are seen by light reflected by them from the Sun and do not emit light like stars. They are, in order starting closest to the Sun, Mercury, Venus, Earth, Mars, Jupiter, Saturn, Uranus, Neptune, Pluto. The four closest to the Sun are small, dense and rocky with an iron and nickel core. The others are larger (except for Pluto) and are icy and less dense. **Except for Pluto the orbits of the planets all lie in the same plane.

The planets
(not to scale)

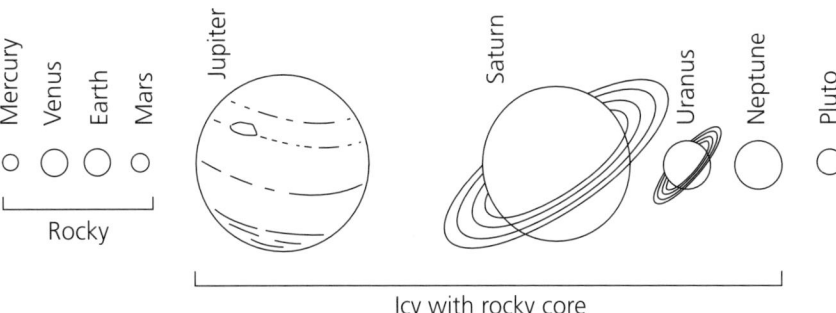

Rocky

Icy with rocky core

- **Moons** move in orbits round planets. Some planets have several moons, some have none at all, the Earth only has one.
- In order to stay in **orbit** at a particular distance, planets, satellites and moons move at a particular speed. An object takes longer to go round its orbit if it is further away from the larger object at the centre. Some satellites, like the Earth, also spin on their own axis as they go round the orbit.
- ****Comets** are smaller, icy objects that are in orbits that are not circular. Sometimes they are close to the Sun and then swing much further away. The orbit of a comet will be in a different plane to the plane containing the orbits of the planets.
- **The **asteroids** are pieces of rock in a belt with an orbit between Mars and Jupiter.

Galaxy facts

- The stars appear to be in fixed patterns called **constellations**. We can recognize these patterns and it helps us to find particular stars. The stars in constellations are often not really connected at all and are huge distances apart – they just happen to be in those particular directions. The pole star (**Polaris**) seems to stay in the same place because it is in the same direction as the Earth's axis. The other stars seem to rotate round it as the Earth spins on its axis. The planets move slowly across these fixed patterns and exactly where we see them depends on where they and the Earth are in their orbits.
- The distances between stars and galaxies are so great that the normal units such as km are far too small. A unit called the light year is used instead. A **light year** is the distance travelled by light in one year. Remember that if a star is 100 light years away you will also be seeing it as it was 100 years ago!
- The **universe** is made up of at least a hundred million **galaxies**. Each galaxy is a collection of about a hundred million **stars**. The **Sun**, at the centre of the solar system, belongs to a galaxy called the **Milky Way**. A cloud of gas and dust in space is called a **nebula**.
- Stars are formed when this dust and gas from space is attracted together by gravity. Eventually so much mass is collected together that the temperature becomes high and the atoms at the centre travel so fast that they can take part in a **fusion reaction.** This nuclear reaction releases enough energy to power the star. Smaller masses may be captured by the star and stay in orbit as planets.
- We believe that all the matter was once concentrated in one place. After the 'Big Bang' this was thrown outwards and the universe is still expanding.
- ** Stars don't remain the same forever. The Sun is a common type called a **main sequence** star. Its present stable state may last for millions of years. It will eventually start to run out of nuclear fuel (hydrogen) and grow hundreds of times bigger into a **red giant**. As the giant dies it will lose some of its matter out to space and become a much smaller **white dwarf**. This too will eventually run out of fuel and become a black dwarf.

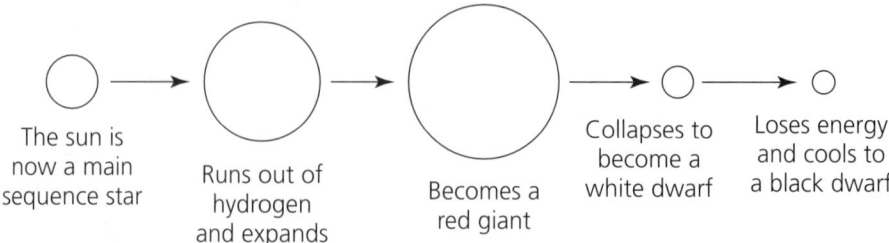

The sun is now a main sequence star   Runs out of hydrogen and expands   Becomes a red giant   Collapses to become a white dwarf   Loses energy and cools to a black dwarf

A star much bigger than the Sun might explode as a **supernova**, throwing most of its mass out into space. The part left behind would be so dense that even light could not escape from its gravity – it would be a **black hole**. During a star's life some lighter elements are joined together to make heavier ones. Some of these are already present in the Sun which means that some of the dust that it was made from originally came from earlier stars that had disintegrated. These heavier elements are also present in the Earth's crust and the other inner planets so it looks as though the planets were also made from the dust of earlier stars.

▶ **The light given out by a star has particular patterns of wavelengths depending on the elements in the star. When we examine the patterns closely we find that the wavelengths have become longer than we expect. The wavelengths have moved towards the red end of the spectrum, so this effect is called **red shift**. We also find that more distant stars have a bigger red shift.

▶ **The '**Big Bang**' theory is an attempt to explain the origin of the universe and also to fit in with other effects such as red shift. The idea is that all the matter in the universe was concentrated in one place and that there was then a huge explosion. All the matter then spread outwards quickly and some of it later condensed to form the stars and galaxies. Since these are all moving outwards they seem to be moving quite quickly away from any observer and this causes the waves of light to be spread out. We see this as the red shift. The more distant galaxies will have a bigger red shift because they are moving away from us faster. There have been many theories to explain our observations of the planets and stars and these change as we discover new facts and try to include them in our theory. (e.g. A few hundred years ago scientists believed that the other planets and the Sun rotated round the Earth, and that the Earth was the centre of the universe.)

▶ ****Rocks** in the crust of the Earth can be **dated** by using the proportions of radioactive materials in the rock. Uranium isotopes with a long half-life decay in a series of much shorter half-life steps to make stable isotopes of lead. The proportions of the isotopes can be used to find the age of the rock and to estimate the age of the Earth. The proportion of radioactive potassium-40 and its decay product argon-40 can also be used in a similar way if the argon gas remains trapped in the rock. Both of these processes give the dates of *igneous* rocks. If you are not sure about words such as isotope or half-life then go through the section on radioactivity.

REVISION ACTIVITY

Before going on to do the practice questions make sure that you can answer all of the following. This will also provide a quick revision test before your examinations.

1 What is the difference between a planet, a star and a galaxy?
2 Name the planets, in order, stating with the one nearest to the Sun.
3 What is the advantage of a geostationary satellite?
4 Name the three types of seismic wave and briefly explain what sort of wave each is and where it can travel.
5 What sort of reaction produces the energy that powers a star?
6 Which galaxy does the Sun belong to?
7 What is meant by a *light year*?
   Why do we need this unit?
8 **What is meant by *red shift*?
   How is red shift explained by the Big Bang theory?
9 What is a *constellation* and what do we use it for?
10 How can rocks in the crust of the Earth be dated?

EXAMINATION QUESTIONS

Questions that are most likely to be on the higher paper only are marked **

Question 1
The force that keeps a satellite in orbit is:
A gravity
B centrifugal force
C friction
D not dependent on the mass of the satellite
E only dependent on the radius of the orbit

Question 2
The length of the day on a planet is caused by:
A how long the planet takes to orbit its star
B how long the planet takes to spin once on its axis
C the tilt of the planet's axis
D the orbit of the planet round its star
E the season of the year

Question 3
Use some of the following words in the list to fill the spaces in the sentences below.

  light years, Sun, star, solar system, towards us, red shift, moons, planets,
         away from us, comets, Milky Way, constellations.

Galaxies are very large and distances to them and across them are measured in
_____. Our own galaxy is called the _____ . We
think that the galaxies are quickly moving _____ and this causes
the light from them to be changed. This change in the light is called
_____ . The Sun which is at the centre of our _____ ,
attracts the _____ and keeps them in orbit.   [3]

Question 4
Use the following table of data to answer the questions below.

Planet	Diameter km	orbital period h	gravity N/kg	rotation period h
Mercury	4 880	2 100	4	1 400
	12 100	5 400	9	5 800
Earth	12 750	8 760	10	24
Mars	6 800	16 500	4	24
Jupiter	142 800	104 000	26	10
Saturn	120 000	258 200	12	10
Uranus	52 800	736 400	9	17
Neptune	48 400	1 444 500	12	20
Pluto	3 000	2 171 200	1	150

(a) The second planet out from the sun is not named in the table.

   Its name is ....   [1]
(b) Which planet is the largest? ....   [1]
(c) How long is a day on Jupiter? ....   [1]
(d) How long is a year on Mars? ....   [1]
(e) If a man of mass 70 kg went to Mars what would he weigh on the planet's
   surface?

   ....

   .... [1]

(f) The force of gravity on the surface of the largest planet, Jupiter, is only 2.6 times as great as that on Earth. If the planet has 318 times as much mass as Earth, why is the gravity force not much bigger?

....

....

.... [2]

(g) What is the relationship between the distance of a planet from the Sun and the time that it takes to make one orbit?

....

.... [1]

(h) Explain why Neptune has a much lower surface temperature than the Earth.

.... [1]

Question 5

(a) The two main types of earthquake are the P waves and the S waves.
    Of **these two waves** which is:
    (i) a longitudinal wave .... [1]

    (ii) a wave that only travels through solids .... [1]

(b) Explain what we have found out from S waves about the interior of the Earth.

....

....

.... [2]

Question 6

(a) What is the main difference between a Moon and a planet?

....

.... [1]

(b) What is a galaxy?

.... [1]

(c) What is a constellation and what do we use it for?

....

.... [2]

Question 7

(a) The Sun is our nearest star. Explain how a star might be formed.

....

.... [1]

(b) The Sun is in a part of its cycle that is sometimes called the *stable period*. Explain what you think is meant by a stable period.

....

.... [1]

(c) What do you think is likely to happen to the Sun when the stable period ends?

....

.... [1]

(d) (i) What is the name of the process by which the Sun produces its energy?

.... [1]

(ii) *Briefly* explain what happens in this process.

....

....

....

.... [1]

6 The electromagnetic spectrum and colour

Topics marked ** are usually on the higher paper and may not be on all syllabuses.

TOPIC OUTLINE

▶ **Electromagnetic waves** are a vibration that is partly electric and partly magnetic.
  – All of these waves can travel through a vacuum and do so at the same speed (300 000 km/s)
  – All of the waves will travel in a *straight line*.
  – When the radiation is *absorbed* it will make the material which absorbs it hotter and can create an alternating current with the same frequency as the radiation.
  – Light is one form of electromagnetic wave. The waves that are longer and shorter than the visible light cannot be seen and have other properties.
  – Shorter waves have higher frequencies and carry more energy.

High frequency						*Low frequency*
Gamma rays	X-rays	Ultra violet	Visible light	Infra red	Microwaves	Radio waves
Short wavelength						*Long wavelength*

▶ **Gamma** radiation sterilizes food and surgical instruments. It also kills cancer cells. It is detected by photographic film or by a Geiger counter. More penetrating than X-rays, it can be used to take 'shadow' pictures of thick metal pipes and castings.
▶ **X-rays** are absorbed more by bones than soft tissue and will affect photographic film. They are used to take shadow pictures of bones and metals.
▶ **Ultraviolet** radiation causes suntan and shows up fluorescent dyes. It carries enough energy to cause skin cancer if the dose is too high.
  Note that ultraviolet, X rays and gamma radiation can all damage or kill cells. They do this by ionizing atoms in the cells or damaging the DNA in the nucleus.
▶ **Light** is the part of the spectrum that we can see. It is used in optical fibre communications which carry more information than electric wires.
▶ **Infra red** is emitted by hot objects (including people). It is felt as heat when absorbed by the skin. It is used in electric fires, toasters and TV remote controls. Detected by photodiodes.
▶ **Microwaves** are used in telephone communications to send signals to and from satellites and aerial dishes because they are not absorbed much by the air. They are strongly absorbed by water molecules so they are used to cook food which usually contains a lot of water molecules. The same absorption process can cause microwaves to damage living cells by the heat released in the water of the cells.
▶ **Radio waves** are used to carry radio and TV signals. The longer wavelengths are reflected from ionized layers in the upper atmosphere and can be diffracted a little around hills.

▶ **Colour** is caused by the way in which the different frequencies of light affect the retina. White light is a combination of all the colours of the spectrum and can be split into those colours by sending it through a triangular glass prism. This happens because the different colours travel at different speeds in the glass and so are refracted by different amounts. Red is refracted least and violet most. The process is called **dispersion**.

The spectrum

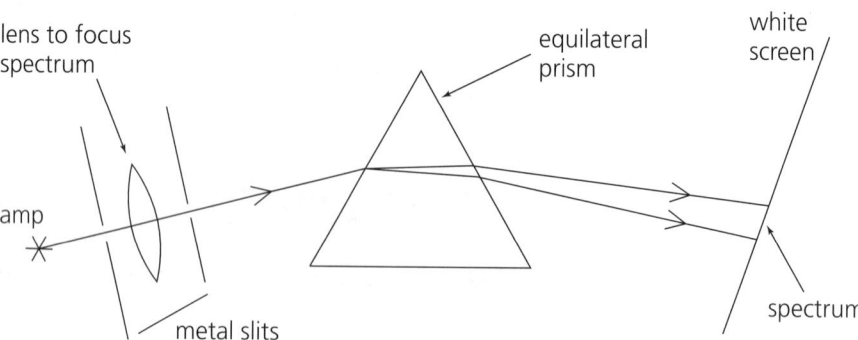

Longest waves   *Shortest waves*

Red   Orange   Yellow   Green   Blue   Indigo   Violet

Remember that objects either give out light, usually because they are hot, or can only reflect the colours of light that fall on them. A red rose can only reflect red light and would therefore appear black in blue light.

▶ The **primary colours** are **red**, **blue** and **green** and cannot be made by *adding* other colours. Remember that these are not the same as in art where the colours are *subtracted* by mixing paint or dye. The three primary colours are used as the colours in TV sets because they can produce other colours when mixed correctly.

▶ When the pure primaries are mixed in pairs they make the **secondary colours**.
  – Red + Blue = Magenta
  – Red + Green = Yellow
  – Green + Blue = Cyan
  – All three mix to produce white.

▶ **Filters** only allow certain colours (the ones that you see) to pass through. They work by *subtracting* colours.
  – A red filter only allows red light to pass through.
  – A cyan filter allows blue and green to pass through.
  – A red filter followed by a blue filter allows no light to pass – black – because there is no colour that can pass through both.
  – A cyan filter followed by blue filter allows blue to be seen as it can get through both filters.
  – Yellow filters are used to darken skies and make clouds stand out better in black and white photography as less blue light gets through the yellow filter.

▶ **The eye**.

The eye

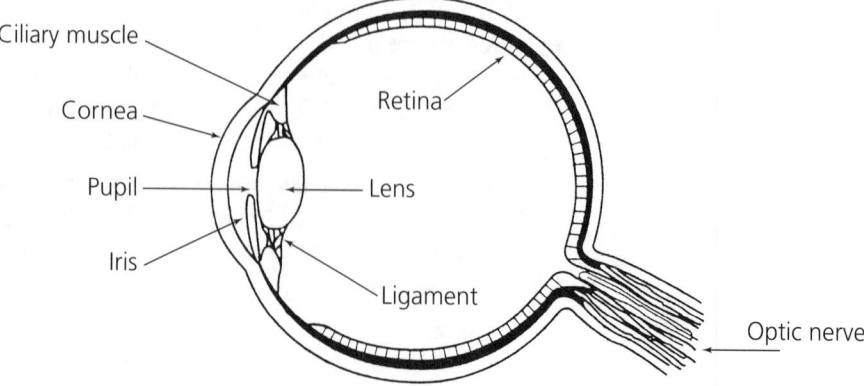

- The **cornea** and the lens focus the light on to the retina.
- The **lens** can be made thinner to focus distant objects and thicker to focus nearer objects. This is done by the **ciliary muscle** which changes the shape of the lens.
- The **iris** is the coloured part at the front of the eye. The hole at its centre that the light passes through is called the **pupil**. The pupil is made smaller in brighter light so that the amount of light reaching the retina remains about the same.
- The image on the **retina** is inverted and smaller than the object. The sensitive surface of the retina reacts to the light falling on it for about $1/16$ of a second and is then replaced by a new picture.
- The information collected by the light sensitive retina is then sent down the **optic nerve** to the brain.
- **Short sight** is caused by the lens being too strong or the eyeball being too long. Distant objects cannot be seen. The correction is by a *diverging* lens.
- **Long sight** is caused by the lens being too weak or the eyeball too short. Near objects cannot be seen. The correction is by a *converging* lens to help the eye lens.

► ****The camera** is similar to the eye – it has a converging lens, makes the same type of image, makes the image on a light sensitive surface and has a diaphragm that does the same job as the iris.

The camera has some differences to the eye – its lens focuses by being moved to and fro instead of changing its shape and the film only takes one picture before needing replacing.

REMEMBER: a lot of the work in the section on Waves, Light and Sound is closely related to this section.

REVISION ACTIVITY

Make sure that you can do the following short checks and then attempt the main revision questions. Remember to write down any equations that you use.

1 Which electromagnetic waves are longer in wavelength than infra red?
2 Red and blue light mix to give _____.
3 The colour that you can see which has the longest wavelength is _____.
4 The colour that you can see which carries most energy is _____ and it has the greatest _____.
5 The electromagnetic wave which gives you sunburn but can cause skin cancer is _____.
6 Short sight is caused by the eyeball being too _____ or the eye lens too _____. The defect can be corrected by a _____ lens.
7 The _____ in a camera does the same job as the iris in the eye. The hole at the centre of the iris is called the _____ and gets _____ in brighter light.
8 Radio waves with a longer wavelength are _____ more than those with a shorter wavelength and this helps them to pass over hills.
9 The three colours used on a TV screen are _____, _____ and _____.
10 The _____ is where the light enters the eye. The light is _____ as it enters and this helps the lens to focus the image.

EXAMINATION QUESTIONS

Questions that are most likely to be on the higher paper only are marked **

Question 1

Radio waves	A	Infra red waves	Visible light	B	X-rays	Gamma rays

Name:

(a) The radiation **A** ....   [1]

(b) The radiation **B** ....   [1]

(c) The radiation with the greatest frequency .... [1]

(d) The radiation that carries most energy .... [1]

(e) The radiation used to kill cancer cells .... [1]

(f) The radiation given out by a hot electric fire .... [1]

(g) A suitable detector for X-rays .... [1]

(h) A use for radiation A..

.... [1]

Question 2

A pupil covers the ends of three torches with coloured filters so that they give out the three primary colours. The beams are then shone on to a white screen so that they overlap as shown in the diagram.

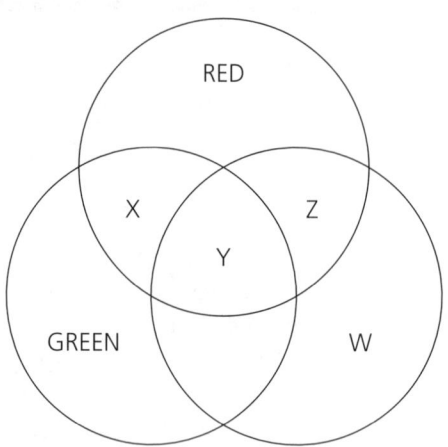

(a) What is the colour W that comes from the third torch? .... [1]

(b) What are the colours produced by the overlapping circles at

(i) X ....   [1]

(ii) Y ....   [1]

(iii) Z ....   [1]

(c) Name one application where primary colours are mixed.

....   [1]

Question 3

(a) Name the part of the electromagnetic spectrum that causes sunburn.

.... [1]

(b) Explain why this sort of radiation can be damaging to the skin and cause skin cancer when the infra red radiation only causes the skin to get hotter.

....

.... [1]

(c) (i) Ultrasound scans are often used to examine pregnant women rather than X-rays. Give a reason for this.

.... [1]

(ii) Explain why we avoid the use of X-rays in the case above but would think that they were safe to use to examine a footballer with a broken leg.

....

.... [1]

Question 4

The diagram shows the path of a beam of white light as it passes through a glass prism.

(a) What is the name given to the change of direction as the light enters the glass at A?

.... [1]

(b) What is the name of the other effect that takes place at A as the light enters the glass?

.... [1]

(c) What colour will be seen at B on the screen? .... [1]

(d) Explain why the colours seen on the screen between B and C appear to be different.

....

....

.... [2]

(e) The white screen is replaced with a blue one. Explain the change in what is seen at C on the screen.

....

.... [2]

Question 5

(a) Name the type of lens that you would find in both a camera and an eye.

.... [1]

(b) Describe the image of a distant object that is formed on the retina of a normal eye.

....

.... [3]

(c) The lens in the eye focuses the light on to the retina. Which other part of the eye is used to help focus the light on to the retina?

.... [1]

(d) Sometimes the lens of the eye is not powerful enough and light from a near object is not focused on to the retina.

(i) This defect is called ....   [1]

(ii) Name the type of lens that will be used in spectacles to correct this problem.

.... [1]

(e) The eye must also be able to focus images of objects at different distances. Explain briefly how the eye does this.

....

....

.... [2]

Thermal energy

Topics marked ** are usually on the higher paper and may not be on all syllabuses. Check carefully whether you need to know all the material under the kinetic theory section and, especially, whether you need to be able to use the equations to do calculations. These may not be on your particular syllabus.

TOPIC OUTLINE

The transfer of thermal energy

You will probably think of **thermal** energy as **heat** energy. The examination boards prefer to call it thermal energy because it is really the total of other sorts of energy such as kinetic energy inside the material.

▶ The thermal energy causes the particles of the material to vibrate. The faster the particles vibrate the higher the energy content and the higher the temperature.

▶ Thermal energy is usually transferred from places at a *higher temperature* to places at *a lower temperature*. There are three main processes by which this can happen.

▶ **Conduction** is the best process through **solids**. The particles at the higher temperature are vibrating quickly and bump into the ones next to them so that the vibration, and therefore the energy, is passed along through the material. This process does not work well with big or complicated particles that cannot vibrate easily. Plastics and most organic materials (e.g. polythene, glass or wool) are bad conductors.

   – **Liquids** are usually poor conductors. You can have ice at the bottom of a boiling tube of water and boil the water at the other end without melting the ice.

   – **Gases** are always poor conductors because their particles are so far apart.

   – **Metals** are *always* very good conductors. **Non-metals** are poor conductors. **This is because metals have free electrons that can carry kinetic energy quickly through the metal.

▶ **Convection** can work when the particles are free to move (i.e. in liquids and gases but NOT in solids). The flow of particles in a convection current then takes the thermal energy with it from one place to another. **When a part of a liquid or gas is heated its particles move faster and it expands, becoming less dense. This less dense, warmer part of the gas or liquid floats upwards and is replaced by more of the cold material. This in turn is heated and a **convection current** is started which carries the thermal energy through the gas or liquid.

Experiments showing conduction and convection

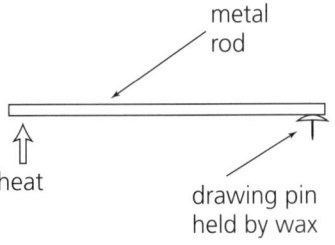

(a) Metals conduct heat

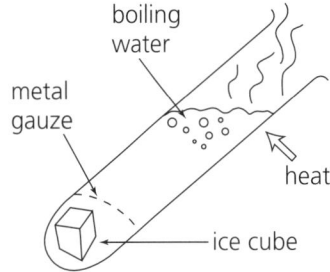

(b) Liquids are poor conductors

▶ **Radiation**, mostly infra red radiation, is emitted by all objects into the space around them. **Hotter objects** will **emit more radiation** than when they are cooler. This sort of radiation can travel through a vacuum – it is how thermal energy reaches the Earth from the Sun. The radiation is **absorbed** by objects that are hit by the waves and the energy has then been transferred from one place to another.

   – The **best emitters** of radiation have dull (**matt**) and **dark** surfaces. The poorest emitters are shiny and light coloured. Matt black is best, shiny silver, like a mirror, is worst.

   – The **best absorbers** are also **matt** and **dark** coloured and the worst are shiny and light coloured. The best absorbers are also the best emitters.

Experiments on radiation

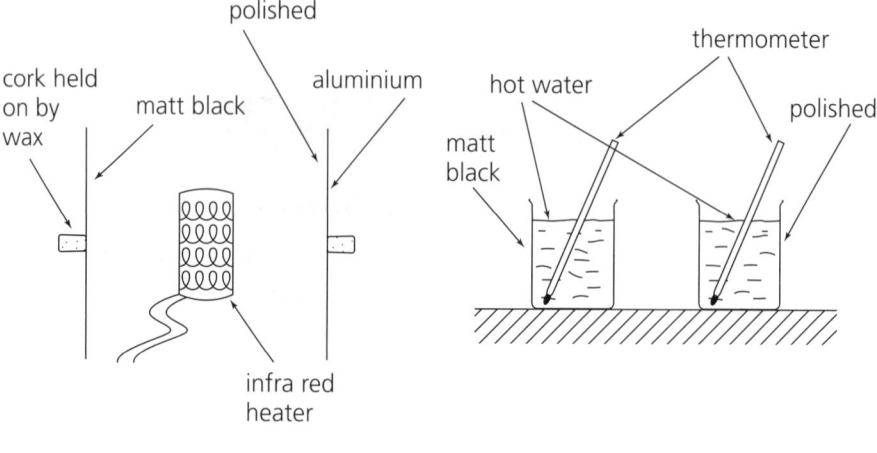

(a) Radiation absorption by identical sized aluminium sheets

(a) Radiation from two identical sized aluminium beakers

   – Some gases, such as carbon dioxide, absorb radiation and do not allow it to pass through easily. This is a cause of the **greenhouse effect** which in turn causes **global warming**. **For the Earth to stay at a constant temperature it must radiate as much energy as it is absorbing. If greenhouse gases absorb some of the radiated energy and then radiate it back towards the Earth then the balance is upset and the average temperature of the Earth rises, changing the climate patterns.

▶ **Insulators** are materials that stop the transfer of thermal energy. Since gases are poor conductors, most insulators have pockets or layers of a gas trapped in them to stop conduction. It is important that the gas is trapped so that it cannot move around and transfer the energy by convection. The material that the gas is trapped in should be a solid that is also a poor conductor. If necessary the outer layer can be silvered to avoid radiation. There are lots of examples, such as layers of clothing, bedding (layers of blankets or duvets full of fibres to trap air).

▶ Avoiding loss of heat energy from houses is important to save energy, resources (and money). Loft insulation, cavity wall insulation and double glazing all trap air, as described in the last section on insulators.

▶ A vacuum flask stops thermal energy going in or out. The vacuum layer stops conduction, the foam or cork stopper stops convection currents and the silvered surface stops radiation.

Kinetic theory

▶ This theory is about what materials are made from and tries to explain why they behave as they do.

   – All materials are made up from *small particles*.

   – The particles in a **solid** all have their own place and can only vibrate. They are very close together. Each piece of the solid keeps its own shape.

    – The particles in a **liquid** are still very close together but are free to move in any direction so that the liquid flows to take the shape of the container.

    – The particles in a **gas** have quite large spaces between them and can move in any direction. The gas can completely fill its container.

    – When the material is given more energy (heat energy) the particles move faster and we notice this as a rise in temperature.

    – When the particles bump into each other or the walls of their container they don't lose energy and don't slow down.

    – As a material is cooled its particles go slower until they stop at the temperature called **absolute zero**. The material then has no internal energy and cannot get any colder. This is the zero of the kelvin temperature scale (see gas laws).

▶ ****Evaporation** happens when some of the faster particles in a liquid reach the surface. The ones going fast enough to get away from the attraction of the others escape into the space above and make a **vapour**. Since the faster particles that escape are the ones with more energy, the liquid is cooled as energy is removed from it. This fact is used in cooling the inside of a refrigerator. **Condensation** is the reverse of this process and heat is then released. The rate of evaporation will depend on the temperature of the liquid (hotter means more fast molecules) the surface area of the liquid (more area means a greater chance of escape) and air movement above the surface (is the vapour carried away or can the molecules go back in again?).

▶ ****Specific heat capacity** is the energy that is needed to raise the temperature of one kilogram of the substance by one degree celsius. It will be measured in J/kg °C. This will mean that the energy needed to raise the temperature of something will depend on its mass, the rise in temperature and its specific heat capacity. Specific heat capacity can be measured by using an electric heater to measure the energy put into some of the substance. If you measure the mass and the rise in temperature of the substance, you can divide them into the energy supplied to find the specific heat capacity. The energy needed to change the temperature of something can be found from the equation shown.

> **KEY POINT**
> *Energy exchanged =*
> *mass × specific heat capacity*
> *× temperature change*

▶ **Changes of state** are the changes between solid and liquid or liquid and gas. Each change of state will happen at a *constant temperature* (the freezing or boiling point for that particular substance) and heat energy will be taken in or given out. This will be shown on a graph showing a cooling curve, where changes of state will be straight lines at constant temperature. The heat energy involved is called latent heat.

Changes of state

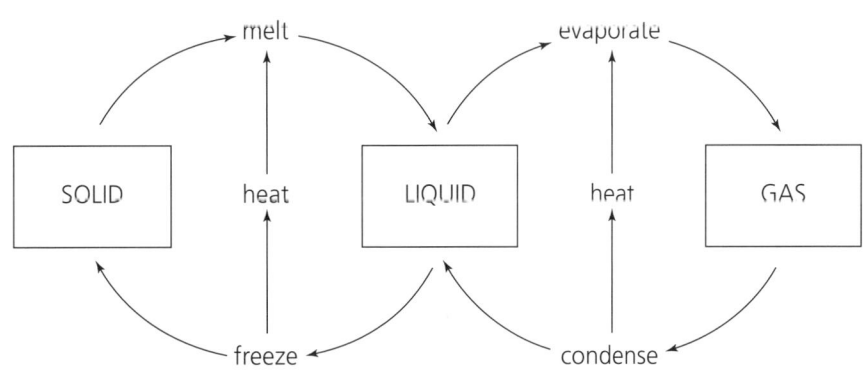

> **KEY POINT**
> *Energy exchanged =*
> *mass × specific latent heat*

**The latent heat that is needed for each one kilogram of the material to change state is called the specific latent heat and will be measured in J/kg. Specific latent heat is greater than specific heat capacity and the latent heat from liquid to gas will be much greater than that from solid to liquid. A scald from steam will be much worse than a burn from boiling water as the large latent heat will be released in addition to the heat released in cooling.

▶ **Kinetic theory and gases**. The pressure of a gas is caused by the collision of its particles with the walls of its container. Each collision pushes on the wall and the total force is caused by huge numbers of collisions so that it seems to be continuous. If the number of molecules stays the same (constant mass) there are three possibilities:

- If the *temperature is constant* then the gas molecules stay at the same speed. If you squash them into a smaller space then they hit the walls more often and the pressure rises. If you halve the space the molecules hit the walls twice as often and the pressure is doubled. This inverse relationship is **Boyle's law**.

- If the temperature rises then the speed of the particles also rises in proportion to the temperature on the kelvin scale. If the *volume stays constant* this will mean that the particles hit the walls more often and the pressure rises. This proportion between absolute temperature and pressure is the **Pressure law.**

- If the temperature rises but the *pressure stays the same* then the gas will expand so that, in the bigger volume, the particles only hit the walls as often as they did before. The volume is therefore proportional to the absolute temperature which is called **Charles' law**.

- Sometimes all three factors change and the connection is then called the gas equation (see the Forces section for more details).

Useful equations

Energy exchanged = mass $\times$ specific heat capacity $\times$ temperature change

$E = mc\Delta T$ (Remember that Δ means 'change in')

Energy exchanged = mass $\times$ specific latent heat

$E = mL$

⭐ **REVISION ACTIVITY**

Make sure that you can do the following short checks and then attempt the main revision questions. Remember to write down any equations that you use.

1 Adding energy to a substance makes its particles go _____.
2 A solid keeps its shape because its particles are only free to

  _____.
3 Gases are _____ conductors. _____ are always good
  conductors.
4 Liquids and _____ transfer heat by _____.
5 Hot objects will _____ a lot of thermal energy.
6 To make a good insulator you would trap _____ of _____ in a
  solid that is not a metal.
7 Evaporation is when the _____ particles with more energy leave the
  surface of a liquid to form a _____.
8 To radiate heat out most quickly the surface of an engine would be _____

  _____.
9 To keep a casualty warm you might use a 'space blanket' which has a shiny
  metallic surface. This will not _____ a lot of thermal energy.
10 Changes of state occur at constant _____.
11 The temperature when a substance has no internal energy is called

  _____ _____.
12 When you halve the pressure of a gas at constant temperature you will also

  _____ ____ _____.

EXAMINATION QUESTIONS

Questions that are most likely to be on the higher paper only are marked **

Question 1
Which of the following statements about heat transfer is correct?
A Convection is a process that only works well in liquids
B Conduction is a process that works best in non-metal solids
C Radiation is energy that is only emitted from hot objects
D The surfaces that are the best radiators of heat energy are also the best absorbers of heat energy
E Conduction is usually the best method of thermal energy transfer in a liquid

[1]

Question 2
Explain in terms of what happens to the molecules:
(a) why the pressure of a car tyre is greater on a hot day than in the winter

....

.... [2]

(b) what happens when a liquid evaporates

....

....

....

.... [3]

(c) what happens to the energy when a piece of ice melts

....

.... [2]

Question 3
(a) A metal tea pot has a shiny, polished outer surface. What effect does this have on the heat loss from the hot tea to the surroundings?

....

....

.... [2]

(b) Sometimes the tea pot is placed on a cork mat and a 'tea cosy' is put over the tea pot to keep the tea hot for longer.

The tea cosy is made from cotton cloth filled with polyester fibres. Explain how the two additions slow down the heat transfer.
Tea cosy:...

....

.... [1]

Cork mat:..

....

.... [1]

Question 4

**Use this equation to answer the following question.

$$E = mc\Delta T$$

(a) An electric kettle contains 1.5 kg of water at 25 °C. If the specific heat capacity of water is 4200 J/kgK, how much heat energy is needed to raise the temperature of the water to its boiling point?

....

....

....

.... [3]

(b) The power of the kettle is 2 kW. How long does it take to raise the temperature to the boiling point?

....

....

.... [2]

(c) If the kettle does not switch off automatically and the water continues to boil, what happens to
   (i) the temperature of the water

.... [1]

   (ii) the energy that is still being supplied.

.... [1]

Question 5

Baked Alaska is a sweet pudding containing a block of ice cream in its centre.

meringue

ice cream

light sponge base

When the pudding is made, it is put into a very hot oven to cook the meringue quickly. The ice cream does not melt because the meringue and sponge base act as insulators.

(a) What is meant by the word **insulator**?

.... [1]

(b) In making the sponge and meringue, the ingredients are whipped up to include air bubbles. How do these tiny bubbles help to prevent the ice cream from melting?

....

....

.... [2]

(c) Meringue can be cooked by itself in a cool oven for several hours. Explain why baked Alaska cannot be cooked in this way.

....

.... [1]

(d) One day a cook found that there were no sponge bases left. It was suggested that the Baked Alaska could be made using aluminium foil cake containers.

Explain why the ice cream would melt.

....

.... [2]

MEG 1996

8 Work, energy and power

Topics marked ** are usually on the higher paper and may not be on all syllabuses. Quite a lot of this work has equations so that you can do simple calculations. These have also been collected together at the end of the notes so that they are easier to find and check.

TOPIC OUTLINE

Work done and energy transferred

▶ During this part of the syllabus the difference between **mass** and **weight** is often important. You should remember that mass is a measure of the quantity of material that you are measuring, and that it *remains the same* wherever you take it. Weight is the *force* with which the mass is attracted by *gravity*. Weight therefore depends on where the object is, but will remain almost the same as long as it stays on the surface of the Earth. A hammer taken to the Moon would still have the same mass but only about $\frac{1}{6}$ of its weight on Earth (because the Moon is smaller and therefore has a smaller gravitational force). On its way to the Moon the hammer will be weightless for a time.

  – On the surface of the Earth each 1 kilogram of mass will have a weight of about 10 newtons. We say that the **gravitational field strength** is 10 N/kg. This is also the same as saying that objects dropped close to the surface of the Earth will accelerate downwards at 10 m/s². You should remember that all objects would do this if placed in a vacuum so that weight was the only force acting. Objects that do not fall with this acceleration have another force acting – e.g. a feather will fall more slowly because the upward force of air resistance can easily be as large as its small weight. (see the section on Forces for an explanation of 'terminal velocity'.)

▶ Whenever *work is done* the same amount of *energy is transferred*. Both work and energy are therefore measured in **joules (J)**.

▶ When an object is moved by a force it will *accelerate*. The size of the acceleration depends on the *force* and the *mass* of the object. A larger force will produce a bigger acceleration, but a larger mass will make the acceleration smaller unless the force is also increased. See the section on Forces for more detail if you need to.

▶ As a force moves an object it is **transferring some energy** and **work is done**. Moving a larger force through bigger distances does more work. Notice that the time taken to do the work doesn't matter in this calculation, the total amount of work is still the same.

> **KEY POINT**
> weight = mass ×
>   gravitational field strength

> **KEY POINT**
> Work done = force ×
>   distance moved in direction
>   of force

Power

▶ Obviously the time does matter sometimes, and it is important whether the energy is being changed quickly or slowly. The amount of work done (or energy changed) each second is called **power** and a power of 1 joule per second is called 1 **watt (W)**. It doesn't matter what form the energy is in – electrical, mechanical, kinetic etc. – it can still be measured in watts.

> **KEY POINT**
> $Power = \dfrac{work\ done}{time\ taken}$
> or
> $Power = \dfrac{energy\ changed}{time\ taken}$

An example
The equations used are all in the equations section.
A crane lifts a 2000 kg load 4 m upwards.
(i) What does the load weigh?
weight = mass × gravitational field strength
= 2000 × 10 = 20 000 N

(ii) How much work is done?
Work done = force × distance moved in direction of force
= 20 000 × 4 = 80 000 J

(iii) If the lift takes 100 seconds what is the output power of the crane?

$$Power = \frac{energy\ changed}{time\ taken}$$

$$= \frac{80\,000}{100} = 800\ W$$

If the electrical power going into the crane is 2000 W, what is its efficiency?

$$\textbf{Efficiency} = \frac{energy\ transferred}{energy\ supplied} \times 100\ \%$$

$$= \frac{800}{2000} \times 100\% = 40\ \%$$

You can use power as well as energy in the efficiency equation as it is the energy used/supplied in a fixed time.

Energy

▶ Energy is *never* lost or created, it can only be in a different form or place. This is called **conservation of energy**. Unfortunately the energy is sometimes not transferred to where you really want it to go. This can be important as you will probably be paying for all of the energy! The *fraction* of the energy that is *usefully* transferred is measured by working out **efficiency**. The rest of the energy is often lost to the surroundings as heat.

▶ Whenever we use energy it begins the process as more concentrated and ends being more spread out. This is sometimes called **degradation of energy**. It is obvious when fuels are burned; the energy is concentrated in the fuel and finally is spread out in the surroundings.

▶ **Fuels** are concentrated sources of energy. Coal, gas, oil, wood are all fuels and release energy when they are burned. Electricity is a **secondary fuel** because it must be made by using one of the other fuels at a power station. We use it because it is convenient to transport and clean when the customer uses it. Coal, oil, and gas are called **fossil fuels** because they were created in the crust of the Earth over a very long period of time. Once they have been used they cannot be replaced and they are examples of **non-renewable** energy resources. Another example is nuclear fuel which slowly uses up ores of uranium which are mined from the Earth's crust. Non-renewable resources should be used carefully and economically so that they do not run out before other sources are properly established.

   – Wood can be replaced by growing new trees provided that the process is properly managed, and is an example of a **renewable** resource. Other renewable resources are solar power, wind power, wave power, tidal power, hydroelectric power and biomass. Sometimes these are called alternative energy sources.

   – Many of these energy resources *originally get* their *energy* from the *Sun* – e.g. oil, coal, gas, biomass, wind power, solar power, wave power. Others do not and include tidal power (oceans attracted by gravity from the Moon), nuclear power (the ore is mined from the Earth's crust) and geothermal energy (energy released from radioactivity in the Earth's core).

▶ In most **power stations** electricity is produced by burning a fuel to produce heat. The heat then turns water to steam and the steam drives turbines to make the alternators produce electricity. This process can be summed up in a **chain diagram** to show what happens to the energy at each stage.

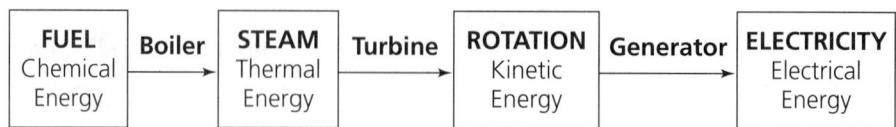

▶ Sometimes an energy change can be summed up by an **energy arrow.** Solar cells, for example, can be used to produce electricity directly.

A solar cell

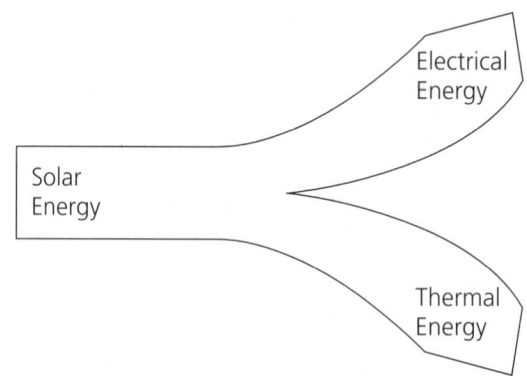

The arrow shows what happens to the energy. Some of it does not change into electrical energy as you would wish. Don't try to make the size of the arrows proportional to the amount of energy.

KEY POINT
Gravitational potential
energy = weight × height
= mass × g × height
or
grav. PE = mgh

▶ **You may be expected to work out the **change in gravitational energy** as an object rises or falls. Remember that, on Earth, *g* will be about 10 N/kg. The value should be given on the examination paper. If experiments at other places (e.g. the surface of the Moon) are described then *g* will be different.

An example
The equations used are in the equations section.
A high jumper has a mass of 65 kg and clears a height of 1.90 m.
(i) What is her change in gravitational potential energy? (*g* = 10 N/kg)
    Change in gravitational potential energy = mass × *g* × change in height
$$= 65 \times 10 \times 1.9$$
$$= 1235 \text{ J}$$

(ii) What happens to this energy as she falls on the other side of the bar?
    The energy will be changing into kinetic energy. After she hits the ground the energy will become thermal energy in the surroundings.

KEY POINT
Kinetic energy =
¹/₂ × mass × velocity²

▶ ** You may be expected to work out the amount of **kinetic energy** that a moving object has. Use the formula carefully and remember that only the velocity is squared. The equation shows the importance of velocity in the amount of kinetic energy that an object has. Doubling the speed will produce four times as much kinetic energy. This is why speed is so important in road traffic accidents.

An example
The equations used are in the equations section.
A space craft has a mass of 1000 kg and is moving at 500 m/s. What is its kinetic energy?
Kinetic energy = ¹/₂ × mass × velocity²
$$= \tfrac{1}{2} \times 1000 \times 500^2$$
$$= 125\,000\,000 \text{ J } (= 125 \text{ MJ })$$

Don't forget about conservation of energy – it may mean that you don't have to work out the kinetic energy. A recent exam question about a high jumper could be done very quickly if you remembered that the PE as she crossed the bar was the same as the kinetic energy when she reached the ground again!

Useful equations

Work done = energy transferred.   *Work and energy units are **joules*** (J)

Force = mass × acceleration. $F = ma$ Force units are newtons (N)

Efficiency = $\dfrac{\text{energy transferred}}{\text{energy supplied}}$ × 100 %

Work done = force × distance moved in direction of force $W = Fd$

Power = $\dfrac{\text{work done}}{\text{time taken}}$ $P = \dfrac{W}{t}$ Power units are watts (W)

Weight = mass × gravitational field strength. $w = mg$ g will be 10 N/kg on
                                                                                 Earth

Gravitational potential energy = weight × height   grav. PE = wh
                                    = mass × g × height   grav. PE = mgh

Kinetic energy = ½ × mass × velocity² $KE = \frac{1}{2}mv^2$

Check your syllabus – you may not need to remember all of these, only to be able to use them. Some of them, especially gravitational and kinetic energy, are usually only needed at higher level in the examination papers and might be given to you.

REVISION ACTIVITY

1 Every moving object has _____ energy.
2 Name a source of energy that does NOT originate from the Sun.
   _____
3 The fossil fuels are _____, _____ and _____.
4 On Earth a mass of 1 kg weighs about _____ and it would still be 1 kg if we took it to the Moon but it would weigh _____.
5 The size of the acceleration produced by a force depends on _____ and _____.
6 A machine does 1000 J of work in 10 s. Work out its power.
7 Energy is always more spread out after we transfer it than before. This is called _____.
8 The kinetic energy of an object depends on its _____ and _____.
9 *Fossil fuels* are *non-renewable*. Explain the meaning of the words in italics.
10 (a) A climber abseils down a 40 m rock face. If his mass is 80 kg what is the change in gravitational potential energy?
    (b) What happens to this energy?

 **EXAMINATION QUESTIONS**

Questions that are most likely to be on the higher paper only are marked **

Question 1
Which of the following energy resources did not originally get its energy from the Sun?

A   Nuclear power
B   Biomass
C   Coal
D   Wind power
E   Solar power   [1]

Question 2
A car has a mass of 1250 kg and is moving along a straight, level road. Its engine is producing a force of 5500 N pushing it forward. There are also counter forces that total 3000 N acting against the car.

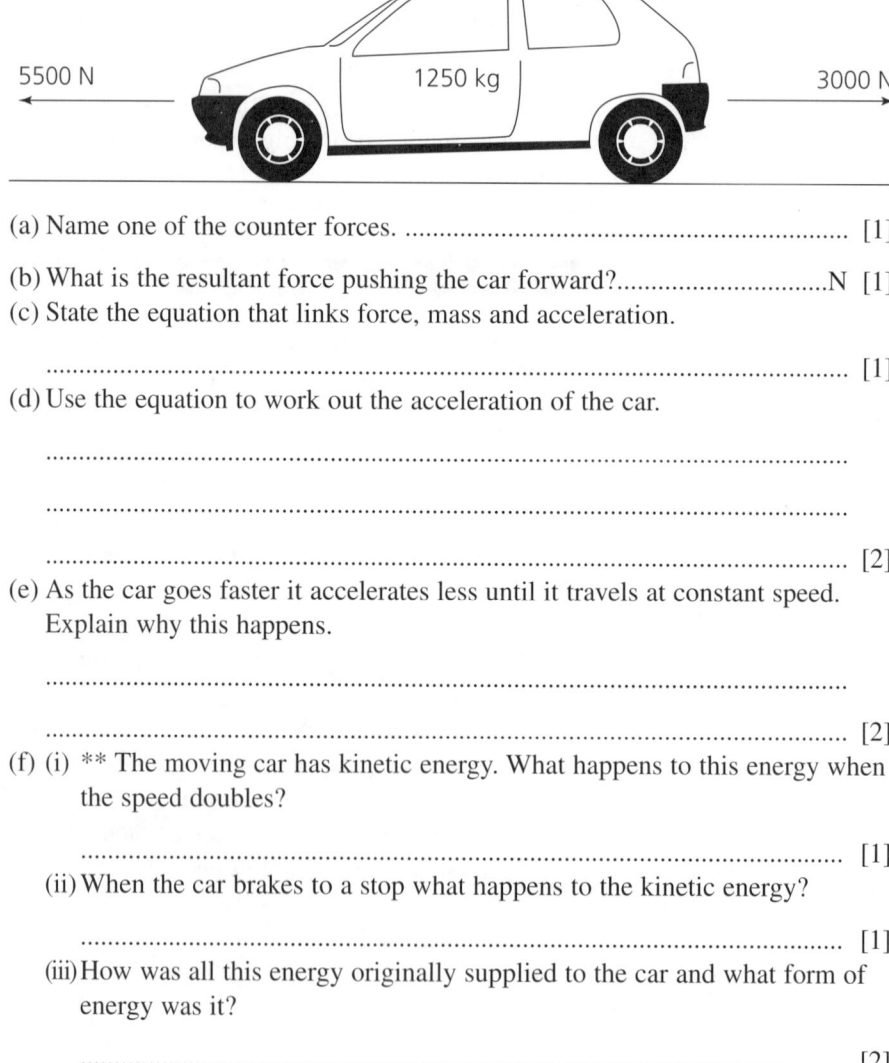

5500 N   1250 kg   3000 N

(a) Name one of the counter forces. .... [1]

(b) What is the resultant force pushing the car forward?..............................N [1]

(c) State the equation that links force, mass and acceleration.

.... [1]

(d) Use the equation to work out the acceleration of the car.

....

....

.... [2]

(e) As the car goes faster it accelerates less until it travels at constant speed. Explain why this happens.

....

.... [2]

(f) (i) ** The moving car has kinetic energy. What happens to this energy when the speed doubles?

.... [1]

(ii) When the car brakes to a stop what happens to the kinetic energy?

.... [1]

(iii) How was all this energy originally supplied to the car and what form of energy was it?

.... [2]

Question 3

(a) Explain clearly the difference between **renewable** and **non-renewable** sources of energy.

....

....

.... [1]

(b) Give two examples of

(i) renewable energy resources ....

.... [1]

(i) non-renewable energy resources ....

.... [1]

(c) Give **two** reasons why it is necessary to cut down the use of non-renewable resources.

(i) ....

.... [1]

(ii) ....

.... [1]

WJEC

Question 4

A crane on a building site is used to lift materials from the ground to the top of the new building. This container of materials weighs 10 000 N and is lifted 25 m at a steady speed.

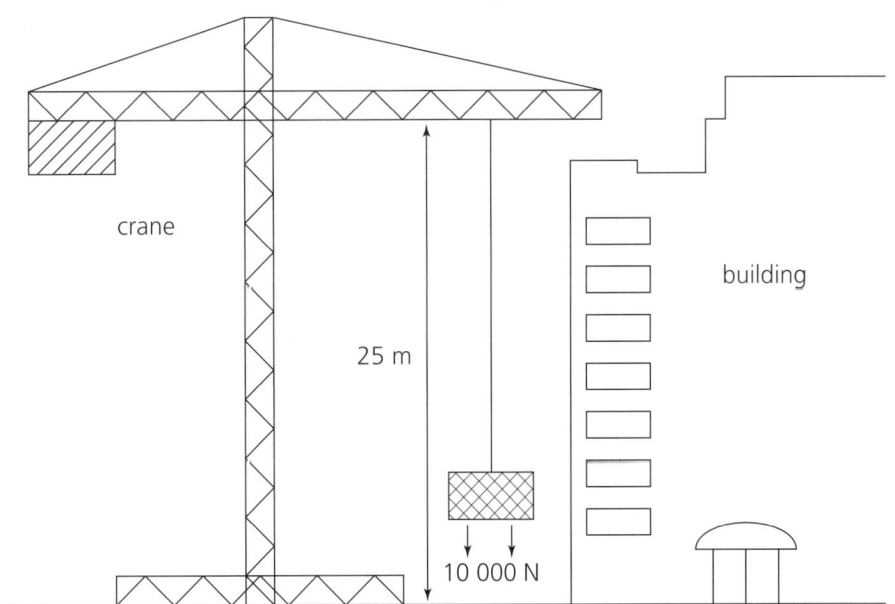

crane

25 m

building

10 000 N

(a) Write down the formula that connects energy transferred, force and distance moved.

.... [1]

(b) Use the formula to work out the energy used in lifting the materials to the top of the building.

....

....

.... [2]

(c) Use the following formula to work out the power produced by the crane if the lift took 50 s.

$$\text{Power} = \frac{\text{energy transferred}}{\text{time taken}}$$

....

....

.... [2]

(d) The electrical energy was actually supplied to the crane at a power of 6 kW.

Use the formula $\text{efficiency} = \frac{\text{power output} \times 100}{\text{power input}}$

to work out the efficiency of the crane.

....

....

.... [1]

(e) (i) Name the type of energy that is gained by the materials as they are lifted.

.... [1]

(ii) Explain what you think happened to the energy that was not usefully transferred to the building materials.

....

.... [1]

9 Radioactivity and X-rays

Topics marked ** are usually on the higher paper and may not be on all syllabuses.

TOPIC OUTLINE

Radioactivity

In a few cases there are two names for some things in this topic. Alternatives are given in brackets. You should try to find out from your particular syllabus which are used.

▶ **Atoms** consist of a central **nucleus** surrounded by electrons in orbitals. The nucleus contains protons and neutrons.
  - The *number of protons* in the nucleus is called the **atomic number** (or the **proton number**) and decides which element the atom belongs to.
  - The *total number of protons and neutrons* in the nucleus is the **atomic mass number** (or the **nucleon number**).
  - There will be as many negative electrons as there are positive protons so that the whole atom is neutral. The following table shows the charge and mass of the particles. The electrons have more mass the faster they go but are always so small that they count as zero mass in an atom.

	mass	charge
proton	1	1
neutron	1	0
electron	0	-1

  It is important to realize that most of an atom is empty space.

▶ All atoms with the same number of protons belong to the same element and will have the same chemistry. Some atoms of the same element have different numbers of neutrons. These are **isotopes** and will have the same proton number but a different nucleon number. Sometimes isotopes have a mixture of protons and neutrons that is not stable or the nucleus has too much energy. The *nucleus* of such an atom will be radioactive and emit radiation. An atom like this is a **radioisotope (radionuclide)**.

▶ If something hits an atom with enough energy, an electron is knocked off. The positive part of the atom that is left behind is called an **ion** and the process is called **ionization**.

▶ Radioactive substances will give out radiation all the time. You cannot turn them off and you cannot alter the speed of the radioactivity no matter what you do. One atom decaying each second is a rate of one **bequerel** (1 **Bq**). There are three main types of radioactivity and all three come from the nuclei of atoms.
  - **Alpha (α)** This radiation is easily absorbed by a few centimetres of air or a thin sheet of paper. Each particle consists of two protons and two neutrons, so it is the same as a helium nucleus. An alpha will travel out in straight lines, knocking electrons off atoms as it goes, until it has lost its energy and stops, leaving a track of ions behind.
  - **Beta (β)** This radiation can pass through a metre or so of air and is absorbed by a few millimetres of aluminium. Each particle is the same as a fast moving electron. A beta will travel out, hitting atoms and causing ioniza-

tion. The track will be longer than for alpha because the small beta can get further between collisions and will not be straight because the particle is easily deflected between collisions.

- **Gamma** (γ) This radiation is not made up of particles at all and is a short burst of electromagnetic waves given out by the nucleus to get rid of some energy. It is *very* penetrating and can pass through many metres of concrete. Several centimetres of lead can reduce it by about a half. It is very short wavelength radiation so it carries a lot of energy. It will travel out in a straight line, occasionally hitting an atom and creating ions.

▶ Radiation from radioactive sources causes **damage to living tissue** by ionizing atoms in the cells. This may kill the cells or it can damage the DNA so that the cell reproduces wrongly, causing tumours and cancers. All doses of radiation can be dangerous but the effect of radiation is cumulative – lots of small doses are as bad as a large dose. The effect depends on the type of radiation and the organs affected. A dose is measured in **sievert**s (**Sv**).

- When the source is outside the body β and γ are most dangerous because they can reach the body whereas the α has too short a range.
- When the source is inside the body α is by far the most dangerous. Its damage will be concentrated in a smaller area because of its range.

▶ **Radioactive materials do not all decay at the same rate. You cannot tell when a particular atom will decay – the process is entirely random – but you can predict what will happen to the billions of atoms in a sample of the material. Half of the atoms will decay in a time called the **half-life**. After two half-lives there will be a quarter of the original atoms left and after a third half-life only an eighth. The half-life for a particular radioisotope can be from a tiny fraction of a second to millions of years. The half-life is very important as it decides the activity (and therefore the danger) of the material. A gram of material with a half-life of a million years may be quite safe, but the same mass of an isotope with a half-life of two hours would be sending out a huge amount of radiation. You need to choose radioisotopes by their activity and their half-life – not by weight!

Plotting the amount of radioactive material left against time always gives the same shape of graph. There are several possible experiments that your teacher might show you which will give the results to plot a graph. You can then find from the graph the time taken to halve the count rate.

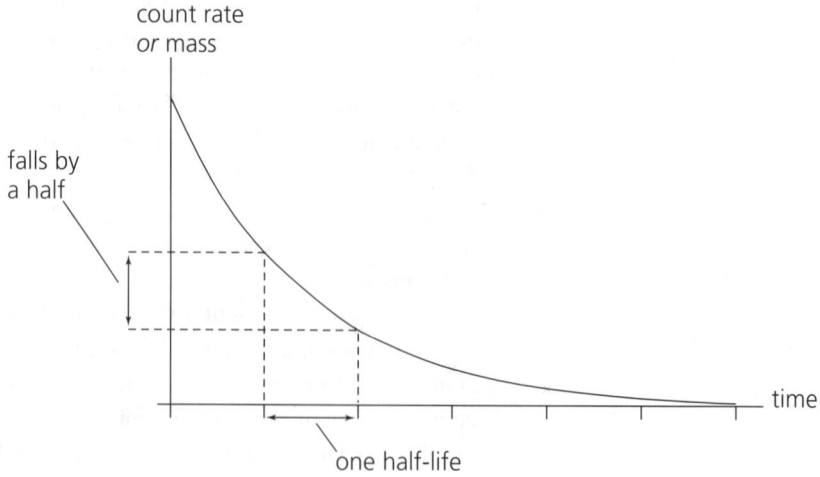

An example

A radioactive source has an activity of 800 Bq and a half-life of 20 mins. How much will be left after one hour?

$$800 \text{ Bq} \xrightarrow{\text{20 mins}} 400 \text{ Bq} \xrightarrow{\text{20 mins}} 200 \text{ Bq} \xrightarrow{\text{20 mins}} 100 \text{ Bq}$$

After one hour the activity is 100 Bq

► There is always radiation around us. This is called **background radiation** and most of it is natural - it is NOT a result of nuclear weapons or nuclear power. You can detect this by switching on a Geiger counter and watching its meter or listening to the clicks from its loudspeaker when there is no artificial source in the room.

There are many sources and you should remember some of the important ones. The percentages will vary quite a lot depending on where you live.
– Building materials 37%
– Cosmic rays 14%
– Food 15%
– Surrounding rocks and ground 20%
– Medical uses 12%
– Discharge from the nuclear industry 0.1%
– Weapons testing, fallout 0.5%
– Miscellaneous such as air travel, luminous dials etc. 0.5%

A lot of the activity from the rocks and building materials is caused by a radioactive gas called radon. In some areas this can collect in caves or even cellars in buildings which therefore need to be ventilated. The dose that you receive from background each year will be about 2.5 mSv – more if you live in Cornwall or Derbyshire where the rocks are more radioactive.

► The best **safety precaution** is to increase distance. Alpha and beta may then not reach you and the gamma will be much more spread out. All sources should be handled by a handling tool to keep them at a distance, and sealed sources used instead of gases, liquids or powders whenever possible. Sources should be logged in and out and stored in a locked metal cabinet. Most sources have a specially designed box with a lead lining to cut down the radiation that can escape. The source containers, cabinet and room door should be marked with the radioactivity hazard symbol.

Radioactivity warning symbol

The sources should be of an activity that is as small as possible for the intended use and one particular person should be responsible for the sources. *No* sources should be used or handled by a person under 16 years of age.

► **Uses**
– **Thickness detectors** usually use a beta source on one side of the material and a Geiger counter on the other. The amount of beta radiation reaching the detector depends on the thickness of the material. This can be used to automatically check the thickness of paper or thin metal sheet as it leaves the rollers of the factory machine.
– **Tracers** are small amounts of radioactive material added to normal liquids or chemicals to see where they go. Radioactive phosphorus could be made into plant feed to see where the element goes in the plant. The quantity of radioactive material will be small and have a half-life short enough to make sure that there is little radiation damage. A similar process uses a radioiso-

tope called technetium-99. It is attached to small amounts of sulphur and injected into the body of a patient to see how well the liver is working. If all is well the gamma radiation that it emits will be detected coming evenly from all of the liver. The radioisotope used has a half-life of about 6 hours. Other radioisotopes are used in a similar way to trace other chemicals in the body.

- Large doses of gamma radiation will kill cells and can be used to sterilize medical instruments or to prevent bacteria or fungus growing on foods such as fruits. This is called **sterilization**.

- **Cancer** cells and tumours in places where it is difficult to operate can also be treated by killing the cells with doses of gamma radiation. The patient may be given a dose of a particular radioactive chemical that goes to the correct place (*contamination*) or be *irradiated* by an outside source.

- Radioisotopes with long half-lives can be used to find the age of rocks or to date dead vegetable material such as wood from archaeology sites. **Carbon-14 dating** works by finding the proportion of carbon-14 that is left in samples of wood or other material that was once alive. Knowing the half-life you can then work out when it died.

▶ **Nuclear reactors.** Some atoms have a nucleus that will split up when hit by a neutron and produce two smaller nuclei, two or three more neutrons and some energy. Uranium-235 or plutonium will do this.

$$_0^1 n + {}_{92}^{235}U \longrightarrow 2 \text{ nuclei} + 2 \text{ or } 3 \; _0^1 n + \text{energy}$$

If the mass of the sample is large enough (*critical mass* or greater) then the neutrons hit other nuclei and the reaction continues and grows. This is called a **chain reaction.** The neutrons work best if slowed down by a **moderator** between fuel pins containing the uranium. **Control rods** can be lowered into the reactor to absorb neutrons and slow the reaction down. The heat energy is removed by sending a **coolant** (gas or water under pressure) through the reactor and used to make steam in a boiler. The rest of the process has the same turbines and generators as a gas or coal fired power station.

The problem with generating electricity in this way is that the 2 new nuclei that are produced are often radioactive and may have very long half-lives. This means that the waste products are very difficult to dispose of, or store, safely. The other possible problems are leaks of radioactive material, usually the coolant, and accidentally letting the reaction go too fast so that the core 'melts down' and the reaction becomes uncontrollable.

▶ ****Nuclear Equations**
You will not need to remember all the symbols for unfamiliar elements – they will be given to you on the exam paper if they are needed. Most questions just need you to fill in missing information. Remember that the numbers at the top must total the same on each side of the equation and so must those at the bottom.

- An *alpha emitter* will lose two protons and two neutrons from its nucleus so that it becomes an atom of the element two earlier in the periodic table and four mass units smaller. You can work out the numbers and symbols if you remember that the alpha is the same as a helium nucleus.

$$_{95}^{241}Am \longrightarrow {}_{93}^{237}Np + {}_2^4 He + \text{energy}$$

- A *beta emitter* will lose an electron from its nucleus. It does this by splitting a neutron so that the electron is thrown out and it keeps a proton. Again the numbers on the top and bottom must balance on each side of the equation so the new atom belongs to an element one higher in the periodic table.

$$\,^{90}_{38}\text{Sr} \longrightarrow \,^{90}_{39}\text{Y} + \,^{0}_{-1}\beta + \text{energy}$$

- A *gamma emitter* gives out a short burst of electromagnetic waves and therefore loses some energy. This often happens after alpha decay. Since the waves have no mass and no charge they do not change the element to which the atom belongs – no equation to do!

X-Rays

▶ Electrons are given off by a hot metal cathode and accelerate towards the anode which has a high + voltage. They hit the anode and most of their energy appears as heat but about 0.5 % of it appears as X-rays.

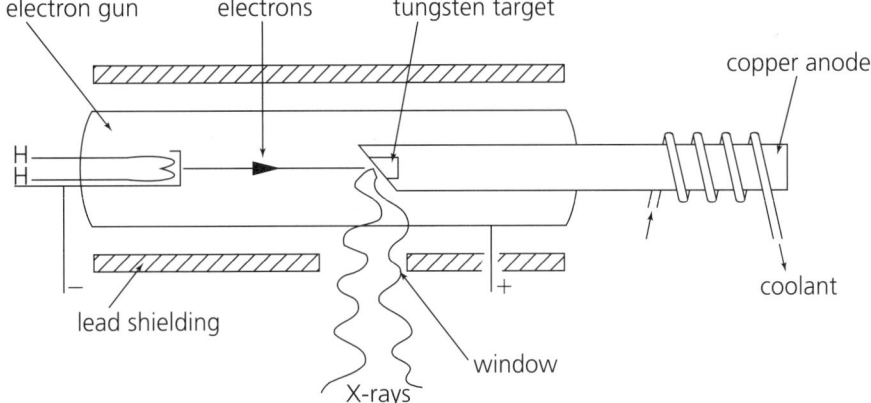

An X-ray tube

This all happens in a vacuum tube that is surrounded by lead shielding to cut down X-rays except though the 'window'. A bigger + voltage on the anode accelerates the electrons more so they go faster and have more energy. They can then make more energetic, more penetrating, X-rays.

▶ The X-rays are absorbed more by bone than by flesh so they make shadow pictures on photographic film producing the familiar 'X-rays' for dental and hospital diagnosis.

▶ X-rays are also dangerous as they can cause ionization and damage cells. The effect is cumulative and repeated doses should be avoided.

Do remember that it is *not* true that all radiation is dangerous. Radiation is any waves or particles spreading out from a source. Both light and sound fit this definition and are not dangerous. The radiation that is dangerous is electromagnetic waves with energy equal to, or greater than, ultraviolet. These waves have enough energy to cause ionization and therefore can damage cells.

REVISION ACTIVITY

Make sure that you can do the following short checks and then attempt the main revision questions.

1 What is the unit for activity?
2 Name the three types of radioactivity.
3 Which of the three types of radioactivity is most penetrating?
4 Where, in the atom, does all radioactivity come from?
5 What is an alpha particle made up from?
6 Name three different types of use for radioactivity.
7 Briefly describe three safety precautions to be taken when using radioactive sources.

8 What are isotopes?

9 What particles does an atom of carbon-14 contain? In what way is it different from its more common isotope carbon-12? Why would a historian be interested in this?

10 ** What is meant by half-life?

11 ** Would X-rays with a shorter wavelength be more or less penetrating?

12 ** Why does a radiographer need to be shielded from X-rays if the dose is small enough to be safe for you?

EXAMINATION QUESTIONS

Questions that are most likely to be on the higher paper only are marked **

Question 1
Isotopes must have:
A The same number of protons as neutrons
B The same number of electrons as neutrons
C The same number of neutrons but a different number of protons
D The same number of protons but a different number of neutrons
E More protons than electrons

Question 2
State two ways in which a radiographer, working with X-rays, can reduce the dangers.

1. .... [1]

2. .... [1]

Question 3
An atom of a radioactive *isotope* has the symbol $^{241}_{95}\text{Am}$
(a) What is meant by an isotope?

.... [1]

(b) How many protons does the nucleus of one atom of the isotope contain?

.... [1]

(c) (i) Which other particles are also contained in the nucleus of the atom?

.... [1]

(ii) How many of these particles are there in the nucleus?

.... [1]

(d) The *half-life* of the isotope is 460 years.
    (i) What is meant by half-life?

    ....

    .... [1]

    (ii) A sample of this isotope is often kept by schools as an example of a material that emits alpha particles. Explain why the half-life makes it suitable.

    .... [1]

Question 4

The diagrams show two radiations being emitted by a radioactive substance and their absorption. Name the radiation emitted in each case.

(a)

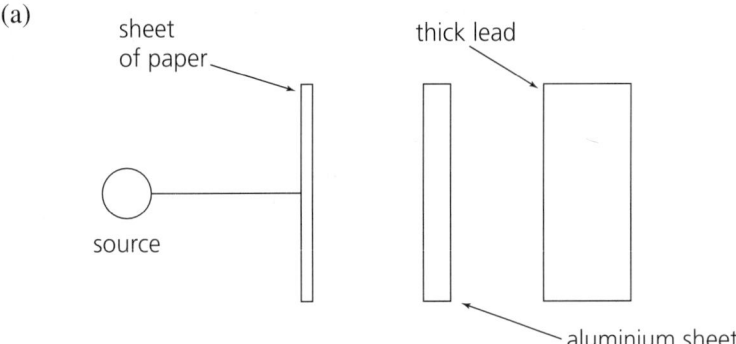

Radiation = ... [1]

(b)

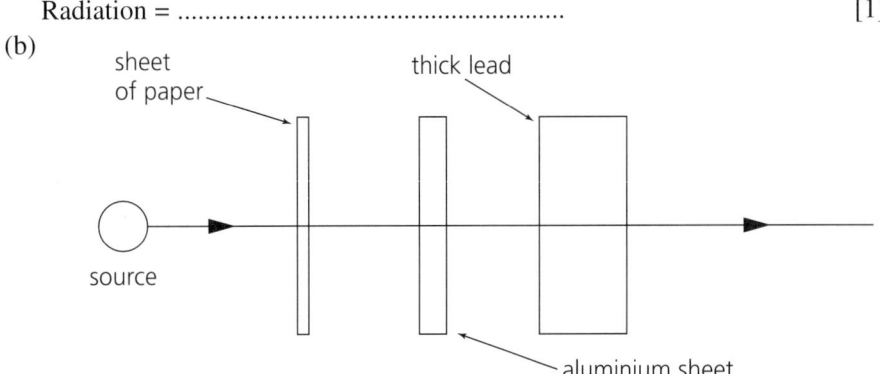

Radiation = ... [1]

Question 5

**A small sample of a radioactive material was collected and tested in a laboratory. It was then tested again by putting a Geiger counter close to it. When first tested the count rate from the sample was 1000 Bq and it fell to 250 Bq after 4 minutes.

(a) What is the half-life of the substance?

....

.... [2]

(b) After a few hours the count rate had fallen to 40 counts per minute and remained constant. What caused the constant reading?

.... [1]

Question 6

There is always radiation around us called *background radiation*.

(a) Name two possible sources of this radiation.

1. .... [1]

2. .... [1]

(b) Give a reason why this background radiation depends on where you are in the country and is not always the same.

.... [1]

Question 7

**The equation below shows a reaction that takes place in the core of a nuclear reactor.

$$^{235}_{92}U + ^{1}_{0}n \longrightarrow ^{236}_{92}U \longrightarrow ^{95}_{39}Y + ^{138}_{53}I + 3^{1}_{0}n$$

(a) State the name of the reaction that is taking place.

.... [1]

(b) What long term problem may be caused by the waste products of this reaction?

....

.... [1]

(c) What is the difference between this type of reaction and the one that powers the Sun?

.... [1]

Question 8

There are a number of uses for radioactive materials.

(a) Name or **briefly** describe a medical use for radioactivity:

.... [1]

(b) Name or **briefly** describe an industrial use for radioactive materials:

.... [1]

part III
All the answers

Solutions
Static electricity and basic circuits

★ SOLUTION TO REVISION ACTIVITY

1 electrons
2 larger, spark, earth
3 repel
4 series, current
5 Any two from: more current, more turns, an iron core
6 6
7 24
8 $R_{total} = R_1 + R_2 + R_3$
   $= 6 + 6 + 6 = 18\ \Omega$
9 less, more
10 **Energy in kWh** = power in kW × time in hours.
   $= 0.1 \times 25 = 2.5$ kWh
   Cost = units × cost per unit = $2.5 \times 7 = 17.5$ p

ANSWERS TO EXAMINATION QUESTIONS

Please note that the answers given for all topics are those supplied by the author. In the case of questions supplied by an examination board the answers and hints are still those of the author and the board accepts no responsibility whatsoever for the accuracy or method of working in the answers that are given.
The marks given for each part of the answer are shown in *italics*.

Question 1
E *(1)*

Question 2
(a) $1.5 + 1.5 = 3$ V *(1)*

(b) $I = \dfrac{V}{R}$ *(1)*

$= \dfrac{3}{4} = 0.75$ A *(1)*

Question 3
(a) Remember that 1000 W = 1 kW

Electrical appliance	Power rating/kW	Power rating/W
Television	0.2	200
Electric kettle	2 *(1)*	2000
Food mixer	0.6	600 *(1)*

(b) The television *(1)*   (c) The food mixer *(1)*

Question 4
(a) (i) All the droplets have the same positive charge so they repel each other. *(1)*
    (ii) The charge will be negative.   *(1)*
(b) Uses less insecticide/better coverage on plant/less insecticide on soil to be washed into rivers.   *(any two at 1 mark each)*

Question 5
(a) (i) The wire X will roll towards the right (towards the power supply).   *(1)*
        It will roll faster in the field and then more slowly as it approaches the supply and leaves the field.   *(1)*

HINT
The positive charge will attract negative electrons from earth

HINT
Use Fleming's left hand rule to find the direction of the movement

(ii) A The direction of movement of wire X is also reversed. *(1)*

     B The force is larger so the wire rolls faster. *(1)*

(b) (i) side P moves up *(1)*   (ii) side Q moves down *(1)*

    (both shown by arrows on the diagram)

(c) The coil is on an axle and the forces produce a moment (turning force) on each side of the coil making it rotate about the axle. *(2)*

(d) The arrows will still show that (i) the force on P is upwards, (ii) the force on Q is downwards. *(1 each)*

(e) The coil stops rotating because the forces are now equal and opposite. *(1)*

(f) One of: reverse the current (best answer), reverse the magnetic field. *(1)*

Electricity and electronics

★ SOLUTION TO REVISION ACTIVITY

1 faster movement, stronger magnet   **2** transformers

3 less   **4** cooler

5 capacitor   **6** emitter, collector, base

7 input, control or process, output   **8** bistable, latch

9 see table in text   **10** smaller, the resistance of the LDR gets smaller so it has a smaller share of the voltage

ANSWERS TO EXAMINATION QUESTIONS

The marks given for each part of the answer are shown in *italics*.

Question 1

(a)

Input P	Input Q	Output X	Output Y
0	0	0	1
0	1	1	0
1	0	1	0
1	1	1	0

(2 marks for each column correct)

(b) NOR.

Question 2

(a)

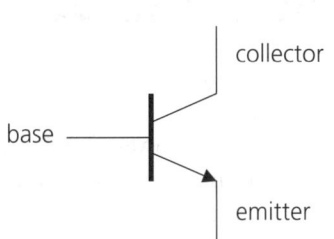

(1 mark for each label – check your spelling!)

(b)(i) The correctly wired circuit is A.   *(1)*

(ii) The bulb in circuit B will not light because the transistor is connected wrongly. (The emitter and collector are interchanged.)   *(1)*

The bulb in circuit C will not light because the battery is connected wrongly. (The voltage is reversed.)   *(1)*

(c) $I_E = I_B + I_C$

$= 2 + 150 = 152$ mA   *(1)*

Question 3

Logic gate X is a NOT.   Logic gate Y is an AND.   *(1 each)*

When the motor is running the light sensor is not in shadow – its output is 0 which then becomes 1 after the NOT gate. The switch is on because the safety guard is down. This makes both inputs to the AND gate = 1 so that its output is on.   *(2)*

If an OR gate was used the motor could run if *either* of the safety devices was set correctly – so it could run with either the safety guard not down or a hand close to the light sensor – and safety is less.   *(2)*

Question 4

(a)(i) AND,   (ii) NOT,   (iii) AND   *(1 mark each)*

(b)

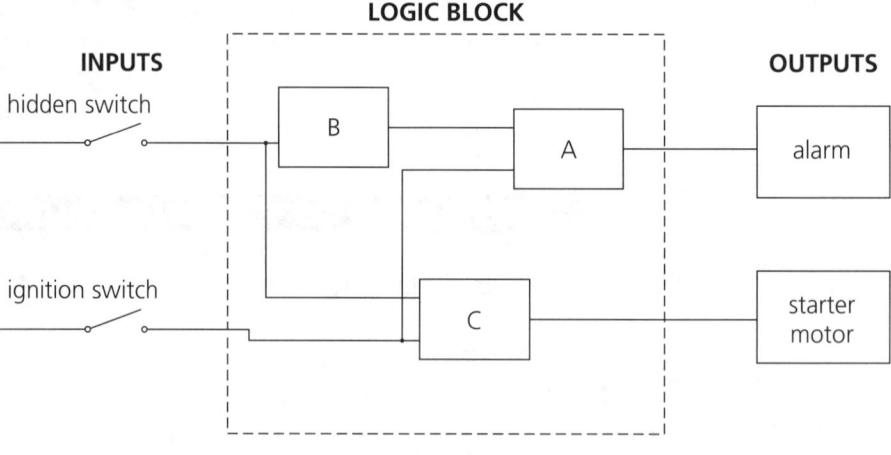

(four wires at $\frac{1}{2}$ *mark each)*

(c) Digital devices have outputs that are either on or off.   *(1)*

> **HINT**
>
> *When showing connections on this sort of diagram remember to show connected wires with dots.*

> **HINT (4c)**
>
> *Not too much detail here – only one mark and a small space for the answer probably means only one fact needed.*

Forces and motion

⭐ **SOLUTION TO REVISION ACTIVITY**

1 12.5 m/s

2 3.3 m/s^2

3 The downward force on the parachutist (weight) will cause the parachutist to begin to accelerate downwards. As the speed increases the air resistance, a force acting upwards, will also increase until it is as big as the downward force. When the forces are equal and opposite there will be no more acceleration and the speed becomes constant – the terminal velocity.

4 A bigger mass or a larger acceleration will both need a larger force.

or: Force = mass × acceleration

5 Several are named in the notes. Most methods make the time for which the force acts longer. This means a smaller acceleration and therefore a smaller force.

6 See the text – the centre of mass should be at the centre of the axle.
7 Up to the *elastic limit* an object is stretched or bent in proportion to the load and it returns to its original size and shape when the load is removed. After the elastic limit the object will become permanently out of shape and the extension is no longer proportional to the load. It is common for materials to stretch more easily after the elastic limit until the breaking point is reached.
8 The amount of gas (its *mass* must stay the same) and its temperature.
9 See the section on equations.
10 The pressure cannot compress the liquid into a smaller space as there are no spaces between the molecules. This means that the same pressure is felt throughout the liquid and the volume remains constant. A gas would compress into a smaller space as the pressure on it was increased.
11 Momentum = mass $\times$ velocity
  = 1200 $\times$ 10
  = 12 000 kgm/s

ANSWERS TO EXAMINATION QUESTIONS

The marks given for each part of the answer are shown in *italics*.

Question 1

acceleration $= \dfrac{\text{change in velocity}}{\text{time taken}}$ *(1)*

$= \dfrac{20}{4} = 5\text{m/s}^2$ *(2)*

Question 2

(a) Pressure $= \dfrac{\text{force}}{\text{area}}$ *(1)*

$100\,000 = \dfrac{\text{force}}{0.6}$

force $= 0.6 \times 100\,000 = 60\,000$ N *(2)*

(b) There will be an equal and opposite force on each side of the glass. *(1)*

HINT
Unit again! All forces are in N so this one is easy to get a mark.

Question 3

(a) (i) air resistance. *(1)*
   (ii) As X increases the acceleration gets less *(1)*
   (iii) The sky diver moves at constant speed when the air resistance is equal and opposite to the weight so the total downward force is zero. *(1)*

(b) (i) acceleration $= \dfrac{\text{change in velocity}}{\text{time taken}}$ *(1)*

$= \dfrac{4 - 0}{0.8}$

$= 5$ m/s^2 *(2)*

   (ii) force = mass $\times$ acceleration *(1)*
    $= 60 \times 5 = 300$ N *(2)*

HINT
Always remember that balanced forces don't change the velocity.

HINT
This question shows the importance of knowing the simple formulas – they get marks themselves and the calculations are often much easier than they look!

Question 4

(a) (i) distance = speed $\times$ time *(1)*
    $= 20 \times 0.6 = 12$ m *(1)*
   (ii) total stopping distance = thinking distance + braking distance
    $= 12 + 30 = 42$ m *(1)*

(iii)

Condition	Thinking distance	Braking distance
Wet roads	remain the same *(1)*	increase *(1)*
Tired driver	increase *(1)*	remain the same *(1)*

(iv) The braking force will be shown by an arrow pointing horizontally
   backwards from the car.   *(1)*

(b) (i) The brake pedal applies a force to the piston in the master cylinder *(1)*
   and creates a pressure in the brake fluid which is transmitted to all the
   wheels *(1)*. The pressure on the brake pads pushes them towards the brake
   disc and the friction between them and the disc creates the braking
   force *(1)*.   *(Total 3)*

(ii) Air is not suitable because it would be compressed into a smaller space
   instead of pushing the pads on to the brake disk.   *(1)*

(iii) pressure $= \dfrac{\text{force}}{\text{area}}$ *(1)*

$= \dfrac{120}{3} = 40 \text{ N/cm}^2$ *(1)*

(iv) The same 40 N/cm^2 will be felt throughout the fluid *(1)* (Reason not asked
   for, but it is because there are no spaces between the molecules of the
   fluid so it can't be squashed into a smaller space.)

(v) force $=$ pressure $\times$ area   *(1)*
   $= 40 \times 50 = 2000$ N   *(1)*

(vi) Kinetic energy in the car *(1)* changes into heat energy in the brakes *(1)*
   and then to heat energy in the surroundings.

4 Waves, light and sound

SOLUTION TO REVISION ACTIVITY

1 unwanted sound
2 energy, vibrate
3 vibrating, vacuum
4 Refraction
5 more, greater, critical
6 $v = f\lambda$, $v = 600 \times 0.5 = 300$ m/s
7 less, shorter
8 concave/diverging, smaller
9 the angle of reflection
10 convex/converging, focus/point
11 focal length
12 See topic notes
13 Telephone/communication cables. Lack of corrosion, less signal loss, greater
   bandwidth (can carry more data at the same time)
14 convex/converging, closer, virtual

ANSWERS TO EXAMINATION QUESTIONS

The marks given for each part of the answer are shown in *italics*.

Question 1
A *(1)*

Question 2

B *(1)*

Question 3

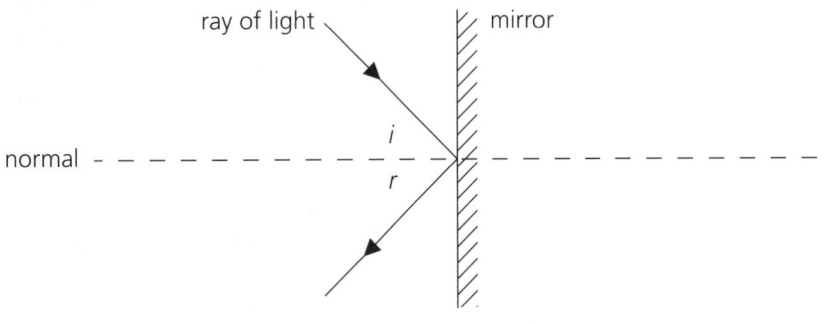

(1 for correct angle for reflected ray)

Question 4

(a)

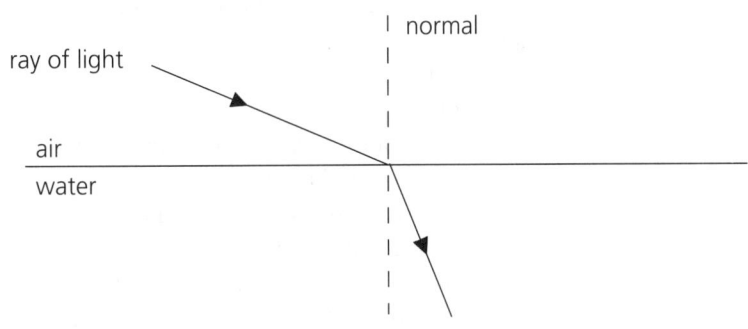

(1 for correct angle for reflected ray)

(b) When the ray enters the water its speed changes – it slows down. *(1)* Since one edge of the ray is slowed before the other the ray is turned as it enters the water. *(1)*

Question 5

The image becomes larger *(1)* and moves further away from the lens. *(1)*

Question 6

(a) speed = frequency × wavelength   *(1)*
            = 2000 × 0.17   *(1)*
            = 340 m/s   *(1)*

(b) speed = frequency × wavelength
    1400 = 2000 × wavelength   *(1)*

$$\text{wavelength} = \frac{1400}{2000} = 0.7 \text{ m}$$ *(2)*

(c) (i) Ultrasonic waves are waves that are like sound waves but with a frequency greater than we can hear.   *(1)*

(ii) distance = speed × time   *(1)*
               = 1400 × 0.1   *(1)* (Use the speed from (b))
               = 140 m   *(1)*

This is the distance there and back so distance to submarine = 70m.   *(1)*

(iii) Examination of pregnant women/industrial/cleaning of delicate instruments etc.   *(1)*

Question 7

(a) In a longitudinal wave the vibration is parallel to the direction of travel. *(1)*
(b) Move the detector from side to side to show that the waves are spread out as they go through the gap and do not only go in the 'straight on' direction. *(1)* The detector will show that the waves are strongest in the straight on direction but there is still a decreasing reading as you move sideways. *(2)*
(c) Greater frequency means that the waves have a shorter wavelength and are therefore diffracted *less* through the same gap. *(1)*

5 *The Earth and space*

SOLUTION TO REVISION ACTIVITY

1 See topic notes
2 Mercury, Venus, Earth, Mars, Jupiter, Saturn, Uranus, Neptune, Pluto
3 Appears to stay in one place above Earth's surface, useful for communications
4 See topic notes
5 A fusion reaction
6 Milky Way
7 The distance travelled by light in one year. To measure huge astronomical distances where km far too small
8 Wavelength of light from stars appears longer than it should – more towards the red end of the spectrum. Galaxies and stars are moving away from each other.
9 A pattern that stars appear to make in the sky which is then given a name. We use it to find the stars that we wish to observe.
10 Rocks can be dated by the proportions of certain radioactive isotopes. See details in the notes.

ANSWERS TO EXAMINATION QUESTIONS

The marks given for each part of the answer are shown in *italics*.

Question 1

A

Question 2

B

Question 3

Galaxies are very large and distances to them and across them are measured in *light years*. Our own galaxy is called the *Milky Way*. We think that the galaxies are quickly moving *away from us* and this causes the light from them to be changed. This change in the light is called *red shift*. The Sun which is at the centre of our *solar system*, attracts the *planets* and keeps them in orbit.

(3 marks = 6 × ½)

Question 4

(a) Venus   (b) Jupiter   (c) 10 h   (d) 16 500 h   *(4 × 1)*
(e) The table says that each 1 kg weighs 4 N on Mars so the man weighs 70 × 4 = 280 N   *(1)*
(f) Because the force of gravity also depends on distance from the centre of the planet and Jupiter is a planet with a large radius.   *(1)*

HINT (4 g)

There is only one mark here for a simple relationship that you can get from the data in the table. If you are interested in astronomy and give a more complex answer such as the square of the orbit period being proportional to the cube of the orbit radius, you won't get additional marks and you might lose them all if you are not absolutely correct – best to stick to the simple answer!

HINT (5b)

Note that the question does say explain and not just state. Don't be afraid to draw a diagram to illustrate this if you can't explain it clearly.

HINT (7c)

You are not asked to discuss the whole cycle for one mark!

HINT (7d(ii))

Note the word 'briefly' and the fact that there are only two marks.

(g) As the distance from the Sun gets greater the time taken for each orbit also gets greater. *(1)*

(h) At the greater distance from the Sun, the surface of the planet receives a lot less energy from the Sun. *(1)*

Question 5

(a) (i) P wave *(1)*   (ii) S wave *(1)*

(b) We know that the Earth has an outer core which is liquid *(1)* because it makes a shadow in the S waves passing through the Earth to the other side. *(1)*

Question 6

(a) A planet is in orbit round a star. A moon is in orbit round a planet. *(1)*

(b) A galaxy is a large collection of stars held together by gravity in one part of the universe. *(1)*

(c) A constellation is a pattern that some stars appear to make in the sky because of their direction from Earth. *(1)* We use it to locate particular stars that are in or close to the pattern. *(1)*

Question 7

(a) A star is formed as dust and debris in space collects together and has enough gravitational pull to collect more and more from the surrounding space until it is large enough to have a high temperature and pressure at its centre. This will then start the fusion reaction that powers the star. *(1)*

(b) A stable period is a time when there is no great change. The Sun is in that period until the fuel for its fusion reaction starts to run out. *(1)*

(c) The Sun will expand to become a red giant. *(1)*

(d) (i) The reaction is a fusion reaction.

(ii) In a fusion reaction the nuclei of small atoms such as hydrogen and helium fuse together to make a larger nucleus. The bigger nucleus needs less energy than the total of the smaller ones, so some energy is released. *(2)*

The electromagnetic spectrum and colour

SOLUTION TO REVISION ACTIVITY

1 microwaves, radio waves
2 magenta
3 red
4 violet, frequency
5 ultraviolet
6 long, strong/thick, concave/diverging
7 diaphragm, pupil, smaller
8 diffracted
9 red, blue, green
10 cornea, refracted

ANSWERS TO EXAMINATION QUESTIONS

The marks given for each part of the answer are shown in *italics*.

Question 1

(a) microwaves   (b) ultraviolet   (c) gamma rays

(d) gamma rays   (e) gamma rays   (f) infra red

(g) photographic film   (h) in microwave ovens/transmitting telephone calls

(All answers 1)

Question 2

(a) blue   *(1)*

(b) (i) yellow *(1)*   (ii) white *(1)*   (iii) magenta *(1)*

(c) television/video recorders   *(1)*

Question 3

(a) ultraviolet

(b) The ultraviolet has a greater frequency and therefore carries more energy. The greater energy is enough to damage cells whereas the energy in infra red is only enough to heat the cells.   *(1)*

(c) (i) The sound waves do not have sufficient energy to damage the cells of the growing baby (especially important while the cells are dividing so rapidly) whereas X-rays can damage the cells, killing or changing them.   *(1)*

 (ii) The damage to an adult, where the cells are dividing much less quickly, is a lot less serious. The dose received when examining a broken leg is too small to be likely to cause any damage to the cells of the footballer.   *(1)*

Question 4

(a) refraction   *(1)*

(b) dispersion   *(1)*

(c) violet   *(1)*

HINT
Remember the red, the longer wave, is refracted least.

(d) The colours all have their own frequency *(1)* and the retina of the eye sees the different frequencies as different colours.   *(1)*

(e) The area will appear to be black because no light is reflected. *(1)* The blue screen can reflect only blue light and it is receiving red light from the prism.
                                                                           (1)

Question 5

(a) convex/converging

(b) erect (the same way up as the object), diminished (smaller than the object), real (can be found on a screen)   *(3 × 1)*

(c) the cornea   *(1)*

(d) (i) long sight *(1)*   (ii) a converging lens *(1)*

HINT
If the lens is not powerful enough then another of the same sort will help it.

(e) The eye can change the focal length of the lens by altering its shape. *(1)* If the ciliary muscle makes the lens thinner it has a longer focal length and focuses objects that are further away. *(1)*

7 Thermal energy

SOLUTION TO REVISION ACTIVITY

1 faster
2 vibrate
3 poor/bad, metals
4 gases, convection
5 radiate
6 layers/pockets, air/gas

7 faster, vapour
8 matt black
9 radiate
10 temperature
11 absolute zero
12 double the volume

ANSWERS TO EXAMINATION QUESTIONS

The marks given for each part of the answer are shown in *italics*.

Question 1
D   *(1)*

Question 2
(a) On a hot day the molecules move faster and collide with the walls more often. This means a greater force on the wall of the tyre and the pressure rises.   *(2)*
(b) The molecules with enough energy are going fast enough to get away from the attraction of the other molecules and leave the surface. They form a vapour above the surface of the liquid. The liquid will be cooled because it has lost the molecules with most energy.   *(3)*
(c) The ice has a crystal structure in which each molecule has its place. The energy is used to break down this structure instead of raising the temperature. The molecules in the liquid water are free and able to flow.   *(2)*

Question 3
(a) The surface will not radiate much heat to the surroundings. Shiny surfaces are poor radiators.   *(2)*
(b) The *tea cosy* is a good insulator because the fibres trap a lot of air. Heat loss by conduction is therefore reduced.   *(1)*
The *cork mat* is also a good insulator and stops heat being transferred to the table through the bottom of the tea pot. *(1)*

Question 4
(a) Temperature rise = 100 − 25 = 75 °C   *(1)*
Energy transferred = $mc\Delta T$
= 1.5 × 4200 × 75 = 472 500 J   *(2)*
(b) Energy transferred = power × time in seconds
472 500 = 2000 × time
time = 472 500 / 2000 = 236 s   *(2)*
(c) (i) it remains the same (at 100 °C)   *(1)*
(ii) the energy is used to change the state from liquid to gas (called latent heat) by separating the molecules to a greater distance from each other.   *(1)*

Question 5
(a) An insulator is a material that is a poor conductor of heat.   *(1)*
(b) Air is a poor conductor and cannot allow convection currents when trapped in small bubbles, so the ice cream is surrounded by an insulator.
(c) The meringue and sponge would slowly become hot and heat energy would then be transferred to the ice cream. The materials are poor conductors but are never perfect and some thermal energy would be transferred if the time was long enough.   *(1)*
(d) The aluminium foil is a metal and so is a good conductor. Heat energy would therefore be conducted through to the ice cream.   *(1)*

HINT (3 a)
The tea also stays hot longer because of this but you are not asked about that in the question

HINT (3 b)
Both these answers are short but must be an explanation. The question does not simply want you to say that the tea stays warm for longer – it tells you that the heat transfer is slowed down and you won't get marks for repeating the question!

HINT (4 a)
This equation will probably be given to you.

HINT (4 b)
Ask yourself if the answer is reasonable. It is just under 4 minutes which is about right and at least indicates that you haven't made a silly mistake.

8 *Work, energy and power*

SOLUTION TO REVISION ACTIVITY

1 kinetic

2 nuclear/geothermal/tidal

3 coal, oil and natural gas

4 10 N, less

5 the size of the force, the mass of the object

6 power = $\dfrac{\text{work done}}{\text{time taken}} = \dfrac{1000}{10}$

    = 100 W

7 degradation of energy

8 mass, speed/velocity

9 see explanations in notes but remember to give *three* answers

10 (a) change in gravitational potential energy = mass $\times$ g $\times$ height

        = 80 $\times$ 10 $\times$ 40 = 32 000 J

(b) The energy is transferred to heat in the metal friction device which slows his decent. (This gets very hot and can permanently damage the rope if he descends too fast!)

ANSWERS TO EXAMINATION QUESTIONS

The marks given for each part of the answer are shown in *italics*.

HINT (2 f)
You need to say more than that the energy increases to get the mark on a higher paper.

HINT (4 a)
Energy transferred will be the same as the mechanical work that is done.

HINT (4 b)
Lots of force and energy problems have marks that you can get easily if you know the unit. In this case you get one mark for joules (J) even if the calculation is wrong – so don't forget to write it down!

Question 1

A   *(1)*

Question 2

(a) air resistance/friction   *(1)*

(b) 5500 − 3000 = 2500 N   *(1)* (No credit for the unit which is given this time.)

(c) force = mass $\times$ acceleration.   *(1)*

(d) 2500 = 1250 $\times$ acceleration

    acceleration = $\dfrac{2500}{1250}$

           = 2 m/s^2 *(1 for size, 1 for correct unit.)*

(e) As it goes faster the counter force of air resistance also increases. *(1)* When it is equal and opposite to the driving force the acceleration will be zero. *(1)*

(f) (i) The energy becomes four times as big   *(1)*

    (ii) The kinetic energy is transferred to heat energy in the brakes.   *(1)*

    (iii) The energy was supplied in the petrol *(1)* as chemical energy *(1)* (OR as energy in the bonds of the chemical).

Question 3

(a) The energy from a renewable source can be used continuously without it running out. This will mean that the source can always be replaced (grow more biomass) or the source itself is almost limitless (solar power).   *(1)*

(b) (i) biomass/geothermal/wind/wave/tide – any *two*   *(1)*

    (ii) coal/oil/gas/nuclear – any *two*   *(1)*

(c) As more are used the resources will eventually run out/some should be left for future generations/the cost will steadily increase – any *two*   *(2)*

Question 4

(a) Energy transferred = force $\times$ distance moved.   *(1)*

(b) Energy transferred = 10 000 $\times$ 25 = 250 000 J   *(2)*

HINT (4 c)
Don't worry that you have just had to use your previous answer to do this. If you got that wrong the examiner won't take off marks again and you can still get full marks for this part. It is fair that you only lose marks once for one mistake!

(c) Power $= \dfrac{250\,000}{50} = 5\,000$ W $\hspace{2cm}$ *(2)*

(d) efficiency $= \dfrac{5000 \times 100}{6000} = 83$ % $\hspace{2cm}$ *(1)*

(e) (i) The materials gain gravitational potential energy. $\hspace{1cm}$ *(1)*

$\hspace{0.5cm}$(ii) It would be used to raise the lift that carried the materials and also as heat energy produced when overcoming the friction forces. $\hspace{0.5cm}$ *(1)*

9 Radioactivity and X-rays

SOLUTION TO REVISION ACTIVITY

1 bequerel, Bq $\hspace{2cm}$ **2** alpha, beta, gamma (α, β, γ)

3 gamma $\hspace{3.5cm}$ **4** the nucleus

5 two protons and two neutrons $\hspace{0.5cm}$ **6** tracers, radiotherapy, thickness monitoring

7 any of those in the text but DO include distance

8 Isotopes are atoms of the same element which have different numbers of neutrons in the nucleus.

9 Carbon-14 has 6 protons and 8 neutrons in the nucleus with 6 orbiting electrons. It is radioactive (a beta emitter with a long half-life) and can be used to date materials such as wood or pieces of clothing.

10 Half-life is the time taken for half of the atoms in a sample of a radioactive material to decay.

11 Shorter X-rays are more penetrating.

12 The dose from X-rays is cumulative so a lot of small doses would be dangerous to the radiographer.

The marks given for each part of the answer are shown in *italics*.

Question 1
D

Question 2
Keep as far away from the machine as possible when it is on (usually in a separate room to operate the controls)/monitor the individual doses and not exceed the maximum dose (the doses are cumulative). $\hspace{1.5cm}$ *(2)*

HINT (3 a)
Since the element is fixed by the number of protons it is exactly the same to say that isotopes are atoms of the same element with different numbers of neutrons.

HINT (3 c)
= 241 − 95

Question 3
(a) Isotopes of an element are atoms with the same number of protons but different numbers of neutrons. $\hspace{1cm}$ *(1)*

(b) 95 $\hspace{6cm}$ *(1)*

(c) (i) neutrons *(1)* $\hspace{3cm}$ (ii) 146 *(1)*

(d) (i) The half-life is the time in which half of the atoms of the material will decay. $\hspace{3cm}$ *(1)*

$\hspace{0.5cm}$(ii) The half-life means that the sample will last for a very long time and that the activity does not change much – it is close to constant for the sort of times that are usual in school experiments. $\hspace{1cm}$ *(1)*

HINT (4 a)
Absorbed by paper.

HINT (4 b)
Some of this even penetrates quite a large thickness of lead.

HINT (5 b)
You don't need to worry about this in part (a) – the question says that the radiation counts in the question are from the sample, so they don't include background.

HINT (7 c)
The question does not ask for details about how the fusion reaction works.

Question 4
(a) alpha *(1)*
(b) gamma *(1)*

Question 5
(a) 1000 ⟶ 500 ⟶ 250
    The count rate has halved twice so the half-life is 2 minutes. *(2)*
(b) The count rate is from background radiation. *(1)*

Question 6
(a) Building materials/food/cosmic radiation/rocks etc. *(any two, 1 each)*
(b) The surrounding rocks and the stone used as building material will have a different radioactivity depending on where you are. *(1)*

Question 7
(a) This is a fission reaction. *(1)*
(b) The products are often radioactive isotopes with long half-lives so that they are dangerous for a long time. *(1)*
(c) The Sun is powered by a fusion reaction where small atomic nuclei are joined instead of large ones being split. *(1)*

Question 8
(a) Killing cancer cells/as a tracer for various body fluids/as a tracer to see if some organs function properly. *(1)*
(b) Measuring thickness of materials/as a tracer in chemical plants/as a sterilizer/to examine thick castings. *(1)*

part IV
Timed papers

Timed practice papers

All candidates should be able to attempt the questions on this paper. A few of you may not need to know the work about electronics – you can omit question 2.
Allow *1 hour* to do this paper.
Attempt *all* the questions.

Question 1
(a) Draw a circuit diagram that shows a battery, two bulbs in parallel and a switch that controls both bulbs. [2]
(b) What is the advantage of putting the bulbs in parallel rather than in series? [1]
(c) What is the name of the piece of equipment that is used to measure electric current? [1]
(d) Name the particles that move round a circuit when a current flows. [1]
(e) Calculate the charge that is transferred in **one minute** when a current of 0.25 A passes through a bulb. [3]
(f) What current is flowing through resistor A in the following circuit? [3]

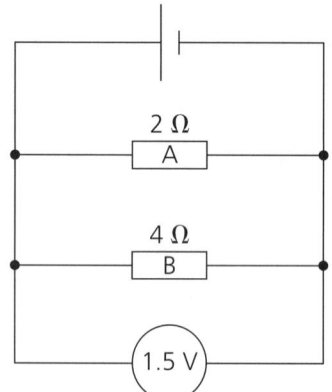

Question 2

AND gate	loudspeaker	OR gate	transistor
LED	LDR	relay	thermistor

Select, **from the list given**,
(a) An input device which detects changes in temperature. [1]
(b) An output device which produces a sound. [1]
(c) A processing/control device which has a high output when either of its inputs are on. [1]

Question 3

A scientist is testing some material and finds that it gives out a form of radiation. When he tests the properties of the radiation he finds that it can penetrate paper but is stopped by a few mm of aluminium. He decides that the material is radioactive. Which sort of radiation is most likely to be being emitted?

Question 4

(a) Each of the following statements describes a solid, liquid or a gas. Read the statements and then write the correct word, *solid*, *liquid* or *gas* next to each one.

The particles vibrate around fixed positions...

The particles spread out to fill all the space available..............................
The particles are close together but slip against each other in all directions.

...........................[3]

(b) The diagram shows air being slowly compressed in a bicycle pump.

Exit closed — Before compression

Exit closed — After compression

Some information about the air inside the pump is given in the table.

	Before compression	After compression
Pressure of air	100 000 Pa	350 000 Pa
Temperature of air	17 °C	17 °C

(i) What is the kelvin temperature of the air inside the pump? [1]
(ii) Using ideas about the movement of particles in gases, explain why the pressure inside the pump increased when the air inside was compressed [3]

SEG

Question 5

(a) The pie-chart shows the distribution of energy use in the UK.

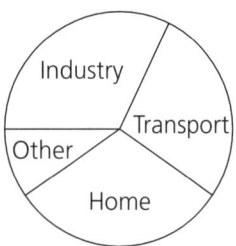

(i) One important use of energy in the home is for cooking. Give **two** other important uses of energy in the home. [2]
(ii) Approximately what percentage of energy is used for transport in the UK? [1]

(b) Most energy needed for transport comes from oil. An alternative is ethanol which is made from sugar cane. In Brazil about 20% of all cars use ethanol fuel.

(i) Oil is a fossil fuel. Name **one** other fossil fuel. [1]
(ii) There are about 10 000 000 cars in Brazil. How many use ethanol fuel? [1]

(iii) Is the energy source for ethanol renewable or non-renewable? Explain your answer. [2]
(iv) Why is it important to use alternative fuels such as ethanol for cars? [1]

SEG

Question 6

(a) The diagram below shows the structure of a **neutral** atom.

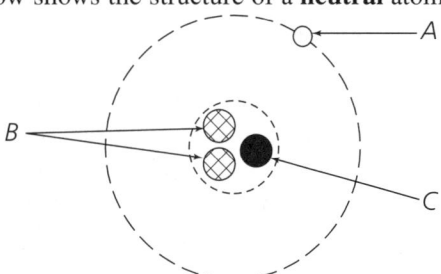

(i) Name the particles marked *A*, *B* and *C*.

A = ... B = ... C = ... [3]

(ii) In what way is the structure of an ion different from that of an atom? [1]

(b) Every atom has a nucleus.

(i) Describe two features of the nucleus of an atom. [2]

(ii) What is radioactivity? [1]

(c) Describe, briefly, two things which happen in the process of nuclear fission.
[2]

(d) (i) Explain what astronomers mean by a solar system. [2]

(ii) Our solar system is part of a galaxy. What is our galaxy called? [1]

(iii) A galaxy is part of a larger system. What is this larger system called? [1]

(e) Figure 1 shows the orbit of a planet around a star.

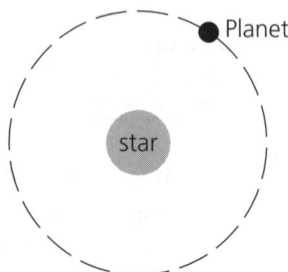

Figure 1

(i) Mark on Figure 1 with an arrow, the direction of the gravitational force on the planet. [1]

Over many millions of years it is possible for a planet to move slowly away from its star, while continuing to orbit, as shown in Figure 2.

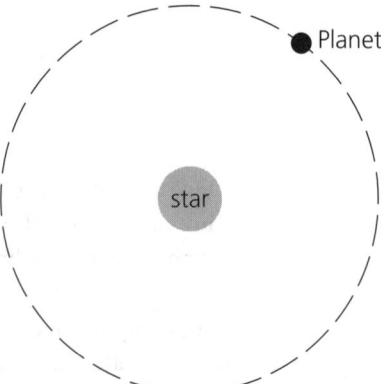

Figure 2

(ii) In what way, if at all, does the size of the force between the planet and star change as the distance between them increases? [1]

(f) (i) Who was the first person to be launched by a rocket into orbit around the earth and what was his nationality? [2]

(ii) Who was the first person to step on to the surface of the Moon in 1969 and what was his nationality? [2]

CCEA

Question 7

(a) In a power station fossil fuel is burned so that the chemical energy in it is transferred into heat energy. Briefly describe the process in which this energy is used to produce electrical energy. [2]

(b) When the electrical energy is produced it is changed to a higher voltage so that it can be transmitted across country. Explain why the electrical energy is transmitted at high voltage. [2]

(c) In one power station the generator produces electricity at 25 000 V. This is stepped up to 400 000 V for transmission on the national grid. The transformer has 3200 turns on its secondary coil.

Use the formula

$$\frac{\text{Voltage across secondary coil}}{\text{Voltage on primary coil}} = \frac{\text{turns on secondary coil}}{\text{turns on primary coil}}$$

to work out the number of turns on the primary coil of the transformer. [2]

(d) The cross country cables could be underground rather than on pylons. Give a reason, other than cost, for using overhead cables. [1]

Question 8

(a) The electromagnetic spectrum contains the following waves. For each one name a **use**.

(i) Infra red

(ii) X-rays

(iii) microwaves

(iv) gamma rays [4 × 1]

(b) Which of the waves named has the longest wavelength? [1]

(c) Which of the waves will carry the greatest energy? [1]

Question 9

In some cases ultrasound is used instead of X-rays in hospitals.

(a) What is ultrasound? [1]

(b) (i) Give an example of when ultrasound might be used in a hospital. [1]

(ii) Why is ultrasound preferred to X-rays in this case? [1]

(c) The ultrasound has a high frequency. Explain why this is an advantage. [1]

Question 10

The diagram shows what happens to the length of a spring when a weight is hung on the end of it.

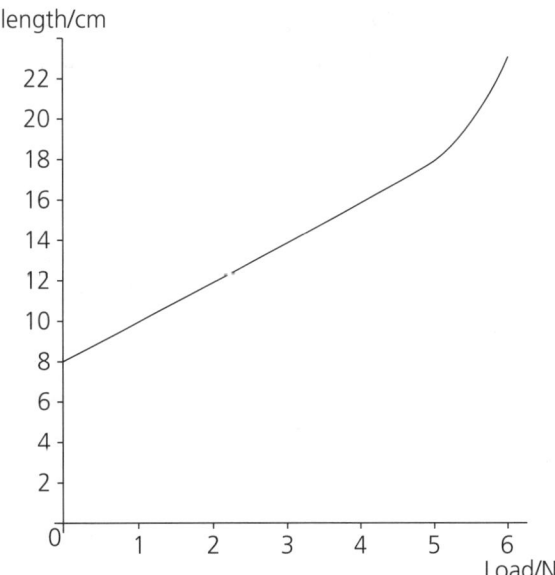

(a) What was the original length of the spring? [1]

(b) How much did the spring stretch when a load of 4 N was hung on it? [1]

(c) Mark on the diagram the part of the graph where the spring is behaving elastically (and obeying Hooke's law). [1]

(d) (i) Use a letter L to label on the graph the point where the spring stops being elastic. [1]

(ii) What is this point usually called? [1]

Question 11

The graph shows the velocity–time graph for a train as it approaches a level crossing. The barrier has not gone down and it automatically does an emergency stop.

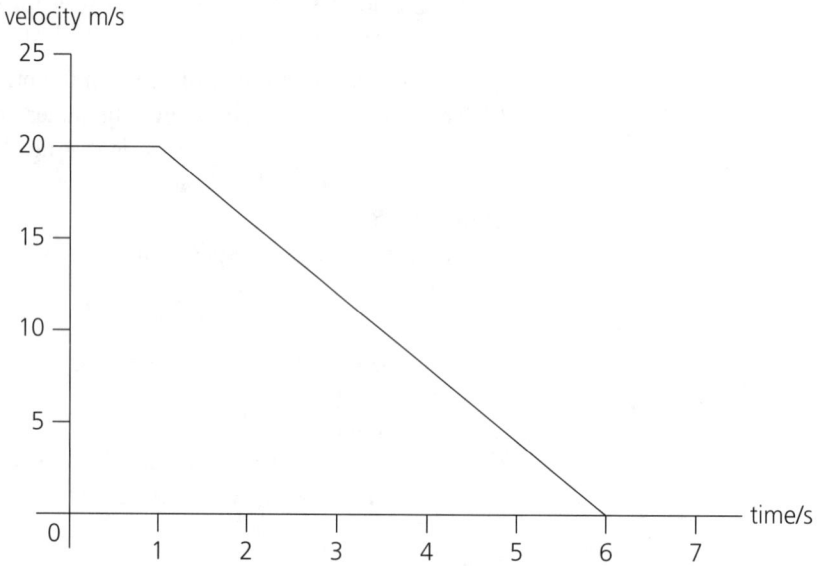

(a) What was the speed of the train before the driver braked?   [1]
(b) How many seconds did it take for the train to stop from that speed?   [1]
(c) What is the acceleration of the train as it brakes?   [3]
(d) What is the distance travelled by the train while it is braking?   [3]
(e) Most level crossings do not have an automatic system like this. Why would
    this make the stopping distance greater?   [1]
(f) If the train has a mass of 80 000 kg, what is the force produced by its brakes?
                                                                                  [3]

Question12

The diagram shows a hydraulic jack.

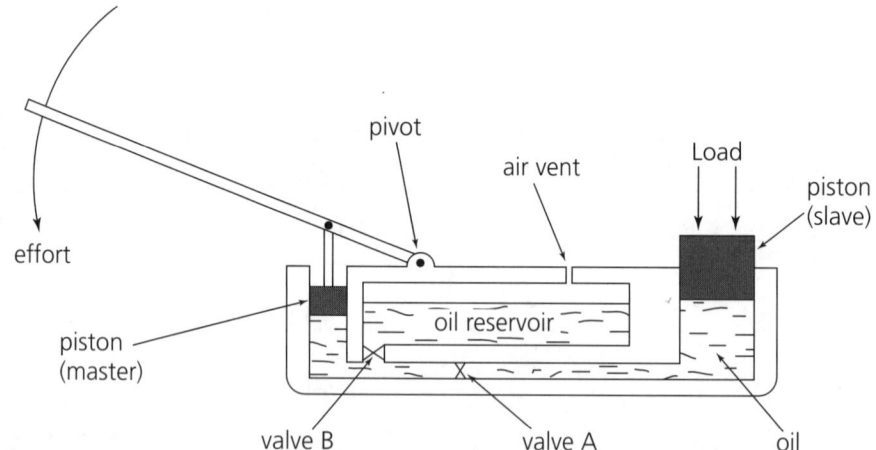

(a) Explain how the hydraulic parts of the jack lift the load. You may find the
    following information about the two valves A and B useful.   [3]

	Valve A	Valve B
Lever going down	open	closed
Lever going up	closed	open

(b) A force of 1000 N is applied to the master piston. If the piston has an area of
    6 cm^2 , what is the pressure acting on it?   [2]

(c) What is the pressure acting on the slave piston at the same time?   [1]
(d) If the slave piston has an area of 30 cm² what is the force acting on the load?
                                                                               [2]
(e) As well as the hydraulic parts the machine also includes another force multi-
    plier. Where is this and how does it help to keep the effort as small as
    possible?   [1]
(f) Why is a gas not used instead of the hydraulic oil?   [1]
(g) Why is water not used as the liquid in the jack?   [1]

TIMED PAPER TWO

This is an additional paper for those of you who are doing the higher paper.
Most questions are suitable for all syllabuses. Omit question 5 if your syllabus
does not include electronics.
Allow *1 hour 15 mins* to do this paper.
Attempt *all* the questions.

Question 1
An electric kettle is used to heat 500 g of water. The graph shows the temperature of the water during the three minutes that the kettle is switched on.

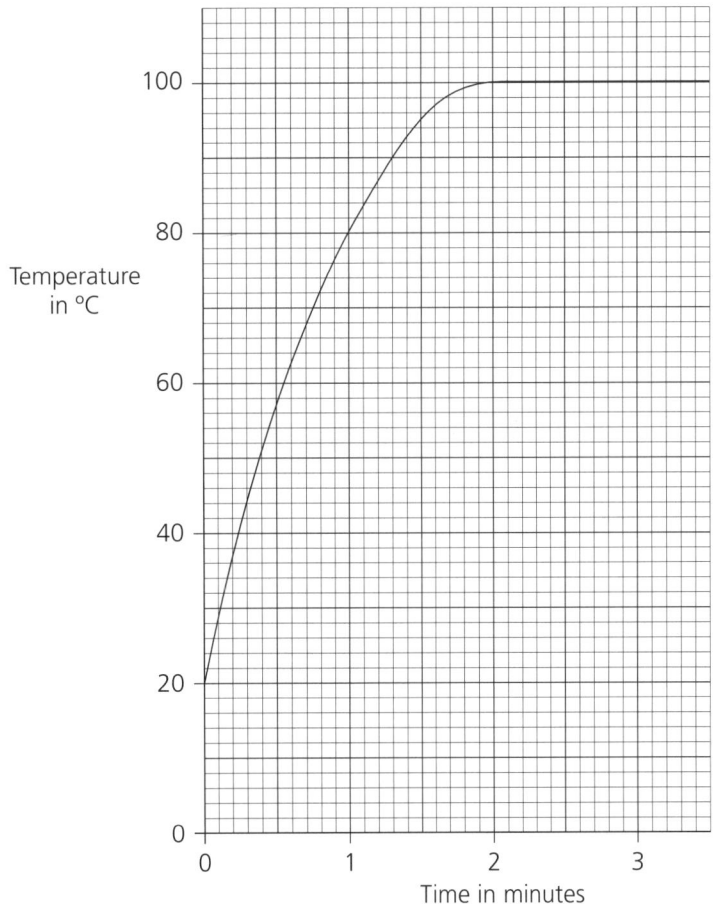

(a) What happened to the temperature of the water during the first minute?   [1]
(b) What happened to the average speed of the water molecules during the first
    minute?   [1]
(c) How long did it take the water to reach boiling point?   [1]

(d) Read the statement below and then cross out the two words that are **not** correct. When the water is boiling, the average speed of the water molecules is

increasing		constant		decreasing	

[1]

(e) Use this equation to answer the following question:

$$E = mc\Delta T$$

When 1 kg of water absorbs 4200 J of energy the temperature of the water goes up by 1 °C.
How much energy does the 500 g of water in the kettle absorb in going from 20 °C to 100 °C?   [3]

SEG

Question 2

The circuit in the diagram can be used to measure the resistance of the resistor marked R.

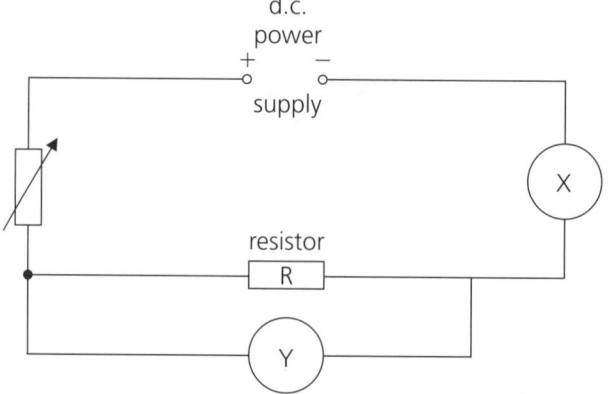

(a) The circuit includes two meters labelled X and Y.
  (i) What is measured by meter X?   [1]
  (ii) What is measured by meter Y?   [1]
(b) What would you do to change the readings on the meters to get other pairs of values?   [1]
(c) The experiment was carried out by a student who plotted his pairs of values on to a graph as shown in the diagram.

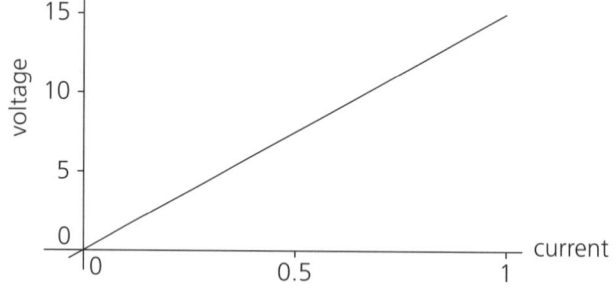

What is the resistance of the resistor? Explain how you got your answer. [2]
(d) How could the student have got negative values to plot on his graph?   [1]
(e) The student then repeated the experiment with a car bulb in place of the resistor. He found that the bulb filament varied from giving out no light to being white hot during his experiment. When he plotted the results on a graph he found that they did not give a straight line.
  (i) Sketch the shape of the graph that he would obtain. [2] (Remember to label the axes)
  (ii) What causes the graph to be this shape?   [1]
(f) A friend suggested that he could investigate another device called a **diode**.
  (i) Draw another sketch to show his results with this new component.   [2]
  (ii) Explain why negative values are important in this case.   [1]

**Question 3

(a) The diagram shows part of the roller coaster opened recently in Blackpool.

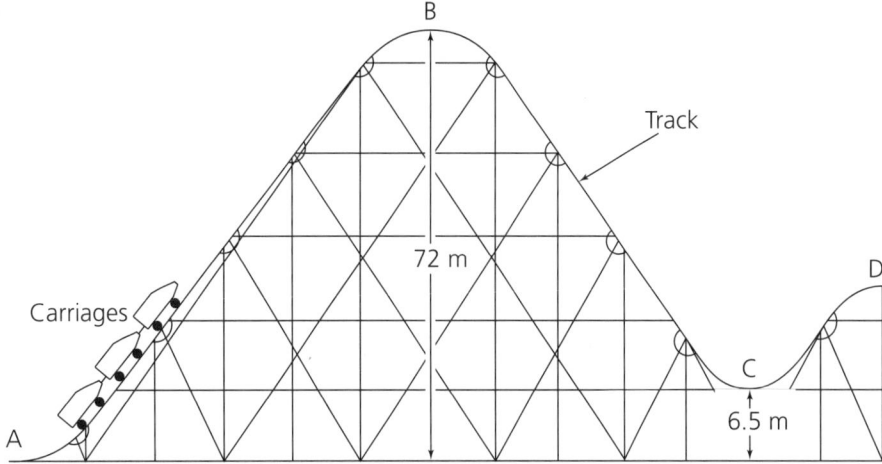

The carriages are pulled up to point **B** by an electric motor. Once a carriage is at point **B**, it is released and free-wheels down the track towards point **C**.

(i) The total mass of carriage and passengers is 3100 kg. How much gravitational potential energy will be gained in moving from point **A** to point **B**? (Take g = 10 N/kg.) [3]

(ii) The power rating of the electric motor is a constant 50 kW. Calculate the time it would take for the carriage and passengers to move from point **A** to point **B**. [3]

(iii) In practice, the time taken to reach point **B** will be longer than you have calculated.
Explain why. [2]

(b) (i) On release from point **B**, the carriage moves down towards point **C**.
Describe the main energy change taking place as it does so. [1]

(ii) By using energy considerations, calculate the maximum possible speed of the carriage as it passes through point **C**. [4]

(iii) Give TWO reasons why the height of the next peak at **D** has to be less than at **B**. [2]

London 1996

**Question 4

The graph below shows how the activity of a radionuclide (radioisotope) changes with time.

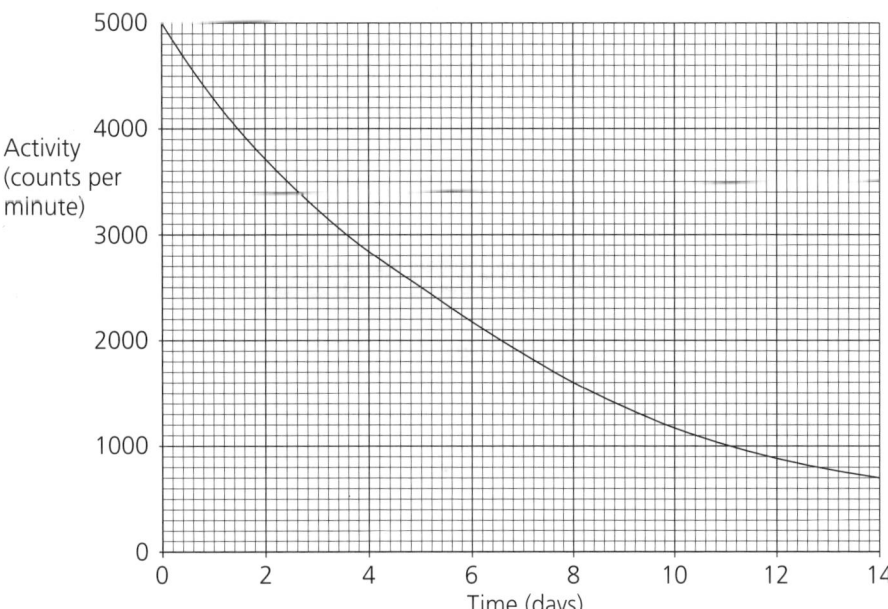

(a) Use the graph to work out the half-life of this radionuclide.   [2]

(b) The radionuclide used to obtain this graph was $^{133}_{54}$Xe.

    (i) How many protons are there in the nucleus of this radionuclide?

    (ii) What other particles are present in the nucleus of this radionuclide?   [2]

(c) $^{133}_{54}$Xe emits gamma radiation.

    (i) State **one** method for detecting gamma radiation.

    (ii) State **two** uses for gamma radiation.   [3]

                                                      NEAB

**Question 5

In this question you may find these equations useful.

$$\text{current} = \frac{\text{p.d.}}{\text{resistance}}$$

p.d. = current $\times$ resistance

A photographic 'dark room' must be kept dark while films are developed. The circuit in the diagram shows a device which warns people not to enter the room when it is dark inside. The LDR is inside the darkroom. The LED is outside the room above the door.

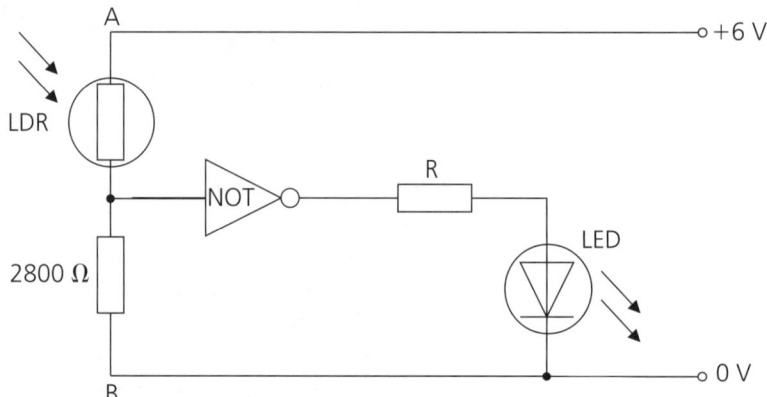

(a) The light in the darkroom is on. The resistance of the LDR is 200 Ω

    (i) Calculate the total resistance between **A** and **B**.   [1]

    (ii) Calculate the current in the LDR.   [3]

    (iii) Calculate the p.d. across the 2800 Ω resistor.   [2]

    The input to the NOT gate is logic 1.

    (iv) What is the logic output of the NOT gate?   [1]

    (v) Is the LED on?   [1]

(b) The darkroom light is now switched **off**.

    (i) What happens to the resistance of the LDR?   [1]

    (ii) What happens to the LED?   [1]

(c) What is the purpose of resistor **R**?   [2]

                                                      MEG

OUTLINE ANSWERS TO TIMED PAPER ONE

Question 1

(a)

HINT

The switch must be in the part of the circuit that controls both bulbs – not in one of the branches

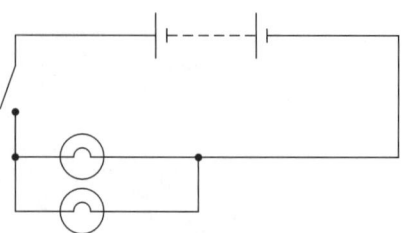

                                                                 (2)

(b) so that if one bulb fails the other still works *(1)*
(c) Ammeter *(1)*
(d) electrons *(1)*
(e) $Q = It$ *(1)*
$= 0.25 \times 60 = 15$ C *(2)*

(f) $I = \dfrac{V}{R}$ *(1)*

$= \dfrac{1.5}{2} = 0.75$ A *(2)*

HINT
All of the voltage (1.5 V) is the resistor A and the parallel resistor B has no effect on what is happening in resistor A.

Question 2
(a) thermistor *(1)*   (b) loudspeaker *(1)*   (c) OR gate *(1)*

Question 3
beta *(1)*

Question 4
(a) solid *(1)*   gas *(1)*   liquid *(1)*
(b) (i) 17 + 273 = 290 K *(1)*
(ii) The smaller volume means that the particles hit the walls more often *(1)* and create a larger force *(1)*. This larger force on the walls creates a greater pressure. *(1)*

Question 5
(a) (i) heating/lighting/washing/TV etc. *(2 × 1)*
(ii) an answer between 25 and 33 % *(1)*
(b) (i) coal, gas, peat etc. *(1)*
(ii) 10 000 000 × 0.2 = 2 000 000 cars *(1)*
(iii) renewable *(1)* more sugar cane can be grown to replace that which is used. *(1)*
(iv) to make limited oil/petrol supplies last longer. *(1)*

Question 6
(a) (i) A = electron   B = neutrons   C = proton *(3 × 1)*
(ii) The number of electrons on an ion is different to the number of protons so it carries an overall charge. *(2)*
(b) (i) at centre of atom/made from protons and neutrons/has a + charge/is small compared to size of whole atom *(any two, 2 × 1)*
(ii) When a particle or electromagnetic radiation is emitted from the nucleus of an atom. *(1)*
(c) nucleus hit by neutron/neutron absorbed/nucleus splits into two/more neutrons released when nucleus splits *(two facts 2, × 1)*
(d) (i) a star and the planets and moons that are held in orbit round it *(2)*
(ii) Milky Way *(1)*
(iii) universe (would also accept 'local group') *(1)*
(e) (i) arrow on planet pointing towards the star *(1)*
(ii) The force decreases with distance. *(1)*
(f) Note that this information is NOT required by all the syllabuses.
Yuri Gagarin – Russian   Neil Armstrong – USA *(4 × 1)*

HINT
Once thermal energy is released to make steam the process is the same, no matter what the source of energy – nuclear, gas etc.

Question 7
(a) The chemical energy is turned into heat energy by burning the fuel and the heat is used to turn water into steam under pressure. The pressure of the steam drives a turbine and that in turn drives a generator. The generator turns the kinetic energy of the turbine into electrical energy. *(2)*

(b) Higher voltage means that the same power is transmitted at a lower current. Lower current means a smaller loss of energy in heating the cables. *(1)*
Give yourself the mark for saying that the energy losses are smaller.

(c) Using the equation given:

$$\frac{400\,000}{25\,000} = \frac{3200}{t}$$

$$t = \frac{3200 \times 25\,000}{400\,000} = 200 \text{ turns} \qquad (2)$$

(d) ease of maintenance *(1)*

Question 8

(a) (i) infra red – TV remote control/radiated heat from electric fires *(1)*
(ii) X-rays – medical uses, detecting bone fractures etc. *(1)*
(iii) microwaves – microwave ovens/transmission of telephone calls *(1)*
(iv) gamma rays – kill cancer cells/examine thick metal castings *(1)*
(b) microwaves *(1)*
(c) gamma rays *(1)*

HINT
Many candidates write gibberish here! A microwave is a wave and it is NOT an oven even though it might be used in one!

Question 9

(a) sound waves with a frequency above the range of human hearing *(1)*
(b) (i) examination of pregnant women *(1)*
(ii) less risk of damage to the rapidly dividing cells of a foetus *(1)*
(c) A greater frequency means a smaller wavelength and this will be diffracted less when it passes through small gaps in the body. The 'picture' is therefore clearer. *(1)*

Question 10

HINT
Read from graph when load = 0

(a) 8 cm *(1)*
(b) 8 cm ($16 - 8 = 8$ cm) *(1)*
(c) The straight part of the graph is the part where Hooke's law is obeyed. *(1)*
(d) (i) Label L where the line on the graph stops being straight (allow anywhere between 5 N and 5.3 N). *(1)*
(ii) elastic limit *(1)*

Question 11

(a) 20 m/s *(1)*
(b) 5 sec *(1)*

(c) acceleration $= \dfrac{\text{change in velocity}}{\text{time taken}}$ *(1)*

$$= \frac{-20}{5} = -4 \text{ m/s}^2 \qquad (2)$$

(d) distance moved   = area under curve *(1)*
$= \frac{1}{2} \times 5 \times 20 = 50$ m *(2)*
(e) Because of the 'thinking distance' of the driver – the train will carry on at its steady speed during the time that it takes the driver to react. *(1)*
(f) Force = mass $\times$ acceleration *(1)*
$= 80\,000 \times 4 = 320\,000$ N *(2)*

HINT
Remember that you can still get full marks for this part even if you got the answer to part (c) wrong. All that matters is that you follow the correct method with the answer that you got.

Question 12

(a) As the master piston is pushed down, the valve at B is closed and the oil is forced through the valve A so that the slave piston is pushed up. At the bottom of the stroke the master piston starts to rise and the valve A closes. Valve B opens and more oil is taken into the master cylinder. At the top of the stroke the piston starts down again and the cycle is repeated. The pressure on the slave cylinder is the same as that on the master cylinder and, because it has a greater area, the force is larger. The distance moved by the slave piston is smaller than that moved by the master piston.   *(3)*

(b) Pressure $= \dfrac{\text{force}}{\text{area}}$ *(1)*

$= \dfrac{1000}{6} = 167 \text{ N/cm}^2$ *(1)*

HINT
Exactly the same pressure everywhere in the liquid!

(c) 167 N/cm²   *(1)*

(d) Force = pressure × area = 167 × 30 = 5000 N   *(2)*

(e) The effort is applied to a *lever* so the effort is smaller than the force that is applied to the master cylinder.   *(1)*

(f) A gas would be compressed into a smaller space and might not move the slave piston.   *(1)*

(g) Water is not usually used because it corrodes the metal of the jack/evaporates quite rapidly/would damage the jack if it froze in the winter. Any of these *scientific* reasons   *(1)*

OUTLINE ANSWERS TO TIMED PAPER TWO

Question 1

(a) The temperature rose from 20 °C to 80 °C   *(1)*

(b) it increased   *(1)*

(c) 2 mins (allow from 1 min 54 s to 2 min)   *(1)*

(d) cross out *increasing* and *decreasing*   *(1)*

(e) $E = Mc\Delta T$

$= 0.5 \times 4200 \times 80$ *(1)*

$= 168\,000 \text{ J}$ *(2)*

Question 2

(a) (i) ammeter – current   *(1)*   (ii) voltmeter – voltage   *(1)*

(b) change the value of the variable resistor   *(1)*

(c) 15 Ω It is the gradient of the straight line on the graph.   *(2)*

HINT
You can also show the working on the graph.

(d) Reverse the connections to the battery so that the current is flowing in the opposite direction.   *(1)*

(e) (i)

HINT
It is always best to draw the graphs with voltage on the 'y axis' if you can so that the gradient depends on the resistance and it is easier to explain. If you swap over the axes the gradient will show conductivity instead.

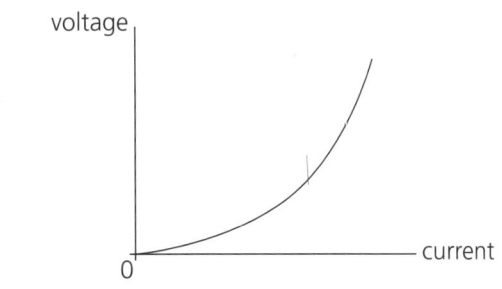

      (2)

(ii) The graph curves because the resistance of the filament in the bulb is increasing with temperature – so the gradient of the graph is increasing.

      (1)

(f) (i)

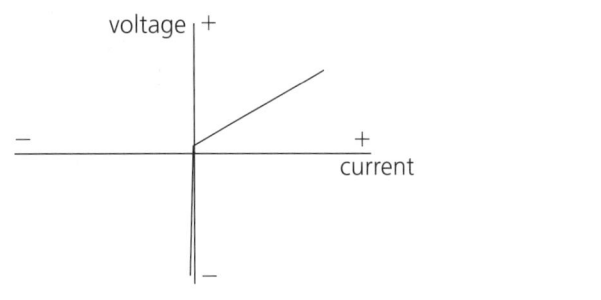

      (2)

(ii) The negative values are needed to show that the diode does not conduct current when it is reverse biased. *(1)*

Question 3

(a) (i) Change in grav. poten. energy = mass $\times$ g $\times$ change in height *(1)*

$$= 3100 \times 10 \times 72$$

$$= 2\,232\,000 \text{ J} \quad (2)$$

(ii) power $= \dfrac{\text{energy changed}}{\text{time taken}}$

time taken $= \dfrac{\text{energy changed}}{\text{power}}$ *(1)*

$$= \dfrac{2\,232\,000}{50\,000} = 44.64 \text{ s} \quad (2)$$

(iii) The time will be longer because energy will also be needed to move against the force of friction. *(2)*

(b) (i) As the carriage moves from A to B the energy change will be from gravitational potential energy to kinetic energy.

(ii) Change of gravitational PE = gain in kinetic energy *(2)*

$$3100 \times 10 \times 65.5 = \tfrac{1}{2} \times 3100 \times v^2$$

$$v^2 = 2 \times 10 \times 65.5$$

$$v = 36.2 \text{ m/s} \quad (2)$$

(iii) The carriage will have a lower speed (and therefore less energy) than just calculated because some will be lost in friction on the way down. The carriage has less energy than would be needed to reach the same height as B (because of the energy used against friction) and some more energy will be lost in friction on the way up to the next peak. *(2 $\times$ 1)*

Question 4

(a) 5 days *(2)*

(b) (i) 54 *(1)*   (ii) neutrons *(1)*

(c) (i) a Geiger tube connected to a rate meter or scalar.

(ii) to kill cancer cells/sterilize equipment or food/examine the interior of thick metal *(Any two, 2 $\times$ 1)*

Question 5

(a) (i) resistors in series add so: total resistance = 200 + 2800 = 3000 Ω *(1)*

(ii) current in LDR is same as current through the 2800 Ω

current $= \dfrac{\text{voltage}}{\text{resistance}}$

$$= \dfrac{6}{3000} = \dfrac{2}{1000} \text{ A} = 2 \text{ mA} \quad (2)$$

(iii) voltage = current $\times$ resistance

$$= 0.002 \times 2800 = 5.60 \text{ V} \quad (2)$$

(iv) logic 0 *(1)*

(v) no, the LED is off *(1)*

(b) (i) the resistance of the LDR increases *(1)*

(ii) the LED is switched on *(1)*

(c) Resistor R limits the current that passes through the diode. *(2)*

HINT

Do this twice, it is the time read from the graph to decay from 5000 to 2500 counts per day which should be the same as the time to decay from 2500 to 1250 counts per day.